DATE DUE			
May 1 '72			
May 15 '72			
Ju 24 '72			
Nov 10 72			
Mar '75			
Nov 7 '75			
Nov 29 '82			
GAYLORD M-2			PRINTED IN U.S.A.

Introduction to
Chemical Thermodynamics

Introduction to
Chemical Thermodynamics
is one of a series of textbooks
published in cooperation
with E. K. Georg Landsberger.

Introduction to
Chemical Thermodynamics

Reuben E. Wood

THE GEORGE WASHINGTON UNIVERSITY

NEW YORK

APPLETON-CENTURY-CROFTS
EDUCATIONAL DIVISION
MEREDITH CORPORATION

5-H.36
W85ν
73657
March, 1971

Preface

Not the least difficult of the decisions I made in writing this book was the decision to use the word *Introduction* in the title. There was, of course, no doubt that the book would not be a comprehensive treatment of chemical thermodynamics. It was intended from the start to be a rather short textbook for use in a one-semester, upper undergraduate, or first-year graduate course of study. The problem is that most if not all of the readers of this book will have been already introduced to a number of aspects of chemical thermodynamics.

I have tried here to set forth the foundations of thermodynamics and to show how a number of principles, especially those useful in chemistry, can be built on these foundations. In some fields, admittedly somewhat arbitrarily selected, I have gone to considerable lengths in developing what approaches the fine structure of the field. This has been done on the premise that even though these detailed developments are only samples of a myriad of such specialized treatments, they do give the flavor of chemical thermodynamics and, hopefully, promote some mental agility that the student can transfer to other problems. It is in the sense of the above purposes that the book is called an introduction.

Because of the differing backgrounds of students in a first course in chemical thermodynamics, one can hope only to minimize the amount of material included in a textbook which is either too elementary for the ablest and most advanced students or which may seem unduly difficult for the least apt or least advanced students. Perhaps this problem is one of the chief reasons that live teachers are still standard equipment in the classroom. I have made an attempt to deal with this problem by literature citations. Perhaps all of the references given as footnotes fall into one of the three catagories: (1) those given just properly to credit the source, (2) those given for the benefit of the student who may need help in an area (mathematical, for example) in which most of the students are already competent, and (3) those given for the benefit of the student who is interested to investigate a subject more extensively than it is treated in this book.

Probably teachers of chemical thermodynamics would agree almost unanimously that few students get a working grasp of the subject without solving a number of problems. In this book problems have been integrated into the text, a procedure that I first met in *Chemical Principles* by Noyes and Sherrill (Macmilian, 1938). Some additional problems are given at the ends of most chapters.

Nomenclature and notation are always problems. I have taken a permissive attitude. Although I have not ignored such recommendations as those of the International Union of Pure and Applied Chemistry, I have intentionally avoided complete uniformity. The student will undoubtedly use the chemical literature and will probably also consult other textbooks. It seems advantageous, therefore, that he should develop some familiarity with commonly-used variants. So, for example, I have used more or less interchangably the symbols μ and $\bar{G}$ to represent the chemical potential, and have tried to make it clear that the same property is commonly called by the various names: Gibbs energy, free energy, Gibbs free energy. One can only guess how far it is wise to go with this. What would be best for one student would probably not be best for another. I have, for example, not used the term free enthalpy which is a perfectly respectable one.

If this book is dedicated at all and if credit is given to those who have helped the author in learning thermodynamics or in the scientific aspects of writing this book, these things will be done in the second edition. Without implicating them in any responsibility for the treatments made, however, three people who gave me much help in preparing the typed manuscript can be mentioned with my thanks. This I do. They are Ava Fowler, Donna Allan, and Suzanne Froman.

R. E. W.

Contents

Introduction to
Chemical Thermodynamics

I

The First Law

ENERGY

Writing on pragmatism William James[1] said,

Metaphysics has usually followed a very primitive kind of quest. You know how men have always hankered after magic, and you know what a great part in magic *words* have always played. If you have his name or the formula of incantation that binds him, you can control the spirit, genie, afrit, or whatever the power may be. Solomon knew the names of all the spirits, and having their names, he held them subject to his will. So the universe has always appeared to the natural mind as a kind of enigma, of which the key must be sought in the shape of some illuminating or power-bringing word or name. That word means the universe's *principle*, and to possess it is after a fashion to possess the universe itself. "God," "Matter," "Reason," "the Absolute," "Energy," are so many solving names. You can rest when you have them. You are at the end of your metaphysical quest.

But if you follow the pragmatic method you cannot look on any such word as closing your quest. You must bring out of each word its practical cash-value, set it at work within the stream of your experience. It appears less as a solution, then, than as a programme for more work and more particularly as an indication of the ways in which existing realities may be *changed*.

Theories thus become instruments, not answers to enigmas in which we can rest.

Energy and its transformations is the grist of thermodynamics. But at the outset, energy is just a word, a name. A proper start in the study of thermodynamics is to identify this name and make of it an instrument.

[1] *Pragmatism*, 1907 p. 43 (Longmans, Green and Co., Inc., New York.) Used by permission of David McKay Company, Inc.

That it is a generic name is obvious from expressions such as kinetic energy, thermal energy, electrical energy, and others. Each of these expressions refers to quantities which can be defined and measured in terms of basic parameters. For example, translational kinetic energy is defined by the equation

$$E_{\text{Trans}} = \frac{mv^2}{2} \tag{1-1}$$

The electrical energy of a capacitor of capacitance C charged to a potential ε is

$$E_{\text{Stat}} = \frac{C\varepsilon^2}{2} \tag{1-2}$$

Gravitational energy:

$$E_{\text{Grav}} = mgh \tag{1-3}$$

The electrical energy corresponding to a current I flowing through a resistance R for the time t is

$$E_{\text{Current}} = I^2 Rt \tag{1-4}$$

For the motion of a body through a distance dx against a force f

$$dE_{\text{Mech}} = f\, dx \tag{1-5}$$

The thermal energy corresponding to a temperature change from T_1 to T_2 of a body whose average heat capacity during the heating is C_P is

$$E_{\text{Therm}} = C_P(T_2 - T_1) \tag{1-6}$$

One may well ask, then, what do all these E's have in common; why are they all called energy? An incomplete answer is that there exist interconvertibility relationships, that E_{Current}, for example, can be transformed into E_{Therm}. But a qualitative interconvertibility is an inadequate discriminator. Momentum is equal to mv. Any body which has a momentum has a nonzero value of E_{Trans} and is capable of producing heat. But momentum is not energy. The criterion is not interconvertibility but a kind of quantitative interconvertibility. Having said this, we have not quite stated but have come close to stating the First Law. In fact, the equation which is called the First Law equation is primarily a definition of energy in terms of measurable components, namely heat and the other particular kinds of energy such as those mentioned above.

James Joule, in a series of experiments carried out during the 1840's established the quantitative relationship between electrical and mechanical energy and the heating effects that these energies could produce. His experiments showed, for example, that when the temperature of a calorimeter (thermally insulated from everything except the heating coil) was made to rise by a passage of an electric current I through a heating coil of resistance R for a time t, the temperature rise from a constant starting temperature would be identical for all experiments in which the value of $I^2 Rt$ was the same even though the values of I and R and t were varied. Likewise, it could be shown that when a moving body of mass m and velocity v was brought to rest by impact or friction in the calorimeter, the temperature rise would be the same for all cases in which the quantity mv^2 had the same value, regardless of variations in the parameters m and v. Moreover, in the case of the electrical heating of an uncomplicated system such as a pail of water it would be found that over small ranges of temperature the temperature rise would be directly proportional to the value of I^2Rt. Thus, a proportionality[2,3] has been shown among I^2Rt, mv^2 and $C_P(T_2 - T_1)$. One is justified, then, to consider these three expressions as different species of the same genus, and the name of the genus energy.

By similar experiments or by logical derivations, other members of the genus could be identified, the pretenders such as the momentum mv could be excluded. Moreover, by establishing a consistent set of units and by doing something about the fact that a given value of I^2Rt will produce different values of $T_2 - T_1$ in a small pail of water and in a large drum of kerosene, i.e., introducing the heat capacity parameter, we can develop expressions for these different manifestations of energy which can be summed to get the total energy change involved in a given experiment.

Before concluding the discussion of the equivalence (which we find to be limited by the Second Law) and additivity of the various forms of energy, two other ideas should be considered. These are the *system* and the *increment*.

The importance in thermodynamic derivations of distinguishing one or more *systems* from the *surroundings* can hardly be overstated. One chooses as a system something of particular interest. The system might be a mole of carbon dioxide—or it might be a steam engine. The surroundings are everything outside of the system or systems; but consideration has to be given only to those parts of the surroundings which interact in some way with the system(s).

Most commonly one deals with closed systems, which are those across whose boundaries matter is not allowed to pass. Less commonly but not infrequently one works with open systems. These can undergo gains or losses of substance. The reason that it is important to specify precisely, even though perhaps quite arbitrarily, the boundary of each system is that one of the most basic and fruitful operations of thermodynamics is the account-keeping of what kind of and how much energy—and in the case of open systems, matter—crosses this boundary during any process which occurs.

---

[2] Nothing in the proposed experiments would tell us that kinetic energy should be written $mv^2/2$ instead of mv^2. The expression $mv^2/2$ comes from Newton's equation $f = ma$ together with the reasonable but nonetheless arbitrary convention that $dw = f\,dx$ and not $2f\,dx$.

[3] At this point C_P may be considered just a property of the system. It will in general vary with the size and composition of the system and, indeed, with the temperature interval. But for a specified system undergoing changes within a specified small temperature interval, it will be a constant.

The bookkeeping is simple. There is a distinction, however, that should be explained. There are many things about systems that can be measured. Their pressures, volumes, temperatures, indexes of refraction, heat capacities, and other characteristics can be measured. Such things are the *properties* of a system. The *state* of a system is described by giving the values of its various properties. However, since these are interrelated, only a few of the properties of a system have to be specified in order to identify uniquely its state. For example, under ordinary conditions a mole of carbon dioxide at 25 °C having a dielectric constant of 1.00700[4] can have only one particular value of each of its other properties including pressure and volume.

PROBLEM 1-1   Both the pressure and the dielectric constant of a gas are functions of temperature and volume. Should the pressure be considered a more fundamental property than the dielectric constant, or should it not? Why? What is the significance of the implied *caveat* "under ordinary conditions" in the statement that "under ordinary conditions a mole of carbon dioxide at 25 °C having a dielectric constant of 1.00700 can have only one particular value of each of its other properties including pressure and volume"? Could an electric or magnetic field affect the properties of the gas?

⠶

When a system undergoes a *change in state*,[5] most of its properties change. The amount by which each property increases is called the increment in the property. If the change is finite, the increment is indicated by Δ. Thus, for example, $T_2 - T_1 = \Delta T$. If the change is infinitesimal, the usual calculus notation, for example dT, is used.

The distinction mentioned above is that between *properties* on the one hand and the quantities heat q and work w on the other. One cannot measure the heat of a system or the work of a system. Heat and work are transients and have significance only during processes. They are the amounts of energy which enter the system across its boundary in a particular form during a particular process.[6] If the same change in state (transition from the same initial to the same final state) were to occur by a different process, the values of q and w could be quite different. An illustration of this fact, which we explore more fully later, is a comparison of two different processes of

---

[4] Referred to some particular frequency.

[5] This is a very general term meaning any change; it includes but is not limited to phase changes.

[6] A partial analogy would be this: A system consists of a solution of sulfur in carbon disulfide. It was made up by adding A grams of rhombic sulfur and B grams of monoclinic sulfur to C grams of carbon disulfide. It is quite correct to say that A grams of rhombic sulfur *went into* the solution and that B grams of monoclinic sulfur *went into* it. But it is meaningless to say that the solution *contains* a certain number of grams of one form and certain number of grams of the other form. In the solution only the total quantity of sulfur can be specified. What form or how much of each form will crystallize out will depend upon the process or processes to which the solution is subjected.

doubling the volume of a gas at constant temperature. One process is to let the gas expand into a vacuum. In this case q is zero or nearly so. If the gas expands against an opposing pressure, q will not be zero or nearly zero.

The contrast, then, is this. Every time the same change in state occurs (as defined by specified initial and final values of a minimum number of properties) every *property* will have the same increment that it had every other time that change occurred even though the change may be carried out one time by one process and other times by different ones. But this is not in general true of *heat* and *work;* except in restricted cases such as we consider in the sections dealing with thermochemistry, there is no nonvariant relation between the initial and final states of a system and the values of q and w which pertain to the process of accomplishing the change from initial to final states. The *heat* and *work* depend on the *process* as well as on the *change in state*. The increments in *properties* depend only on the change in state and must be the same for all conceivable processes of carrying out the same change in state.

The language of mathematics is elegant—and to many, unexplanatory. Increments in properties such as dP, dV, etc. are exact differentials; dq and dw are not. The preceding sentence summarizes a number of the foregoing paragraphs. An equivalent summary is to say that pressure, volume, and others are properties whereas heat and work are not properties of systems.

We can now write a quantitative definition of the amount of energy put into a system. We anticipate our succeeding equations and use ΔE as the symbol for the total amount of energy put into a system during a particular change. The defining equation is

$$\Delta E \equiv q + w \qquad (1\text{-}7)$$

where q is the amount of heat and w the total of all forms of work which enter the system from, or are done on the system by the surroundings.

At this point, in preparation for subsequent examination and use of 1-7, we recapitulate our basis for measurements and evaluations of q and w. Essentially, we start with some particular kind and scale of energy. Probably the classical starting point is the work associated with motion against a force

$$dE_{\text{Mech}} = f\,dx \qquad (1\text{-}5)$$

and a useful secondary measuring stick is the electric energy associated with current flow through a resistance

$$dE_{\text{Current}} = I^2 Rt \qquad (1\text{-}4)$$

Then by definition and comparison we evaluate other forms of energy. The quantitative definition and measurement of heat energy illustrates this method. First, one thermally insulates a system from its surroundings and causes it to undergo some change by the introduction of a known amount of energy— the passage of a measured current for a measured time through a coil of known resistance, for example. Then, after restoring the system to its initial state, the thermal insulation (adiabatic separator) is removed and the system is made to undergo the same change by thermal interaction with the surroundings—by heating with a flame, for example. The definition is, then, that the heat put into the system by the flame is the energy equivalent of the electric work required to produce the same effect in the experiment (adiabatic) in which the system was thermally isolated.

Figure 1-1 represents schematically the kinds of experiments that led to acceptance of the First Law. A system undergoes a change in state from state 1 to state 2. Equation 1-7 together with measurements of q and w

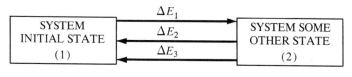

FIGURE 1-1

is used to evaluate ΔE_1. Several different processes are now devised for accomplishing the reverse change in state. The ΔE's for the reverse change are evaluated. The findings are, within the limits of experimental accuracy, (1) that ΔE_2 and ΔE_3 have identical values and (2) that the sums $\Delta E_1 + \Delta E_2$ and $\Delta E_1 + \Delta E_3$ are zero. This result has been found in every case in which such experiments have been done. The First Law of thermodynamics is the extremely useful hypothesis that such results will always be found and that the relationships represent an unqualified natural law.

It is the *unqualified* validity of the defining identity 1-7 for all closed systems together with the assertion that

$$\Delta E = E_{\text{final state}} - E_{\text{initial state}} \qquad (1\text{-}8)[7]$$

that is, that ΔE is dependent only on initial and final states, not on the process of transition, that permits us to view the First Law equation as:

$$\Delta E = q + w \qquad (1\text{-}9)$$

Or we may write the First Law for the infinitesimal process

$$dE = dq + dw \qquad (1\text{-}10)$$

in which the symbol dE represents an exact differential.

From the foregoing, it is clear that for an isolated system, since q and w are zero, ΔE must always be zero. One of the common verbal

[7] Ordinarily one is concerned only with energy differences, and if values are assigned to the energy of a system in a particular state the values are with reference to an arbitrary zero level. These facts do not, however, invalidate the concept of energy as a property.

expressions of the First Law is that the energy of an isolated system remains constant.

Thermodynamics is founded on a few basic articles of faith, and the First Law is one of these. The meaning of that statement and the usefulness of that faith are illustrated in the discovery of the neutrino. Writing in *Nature*[8] three years after they had first detected a free neutrino[9] and twenty-three years after Wolfgang Pauli[10] had first proposed the existence of the particle, Frederick Reines and Clyde Cowan introduce their paper with the following paragraph.

Each new discovery of natural science broadens our knowledge and deepens our understanding of the physical universe, but at times these raise new and even more fundamental questions than those which they answer. Such was the case with the discovery and investigation of the radioactive process termed 'beta decay.' In this process an atomic nucleus spontaneously emits either a negative or a positive electron, and in so doing it becomes a different element with the same mass number but with a nuclear charge different from that of the parent element by one electronic charge. As might be expected, intensive investigation of this interesting alchemy of nature has shed much light on problems concerning the atomic nucleus. A new question arose at the beginning, however, when it was found that accompanying beta decay, there was an unaccountable loss of energy from the decaying nucleus, and that one could do nothing to the apparatus in which the decay occurred to trap this lost energy. One possible explanation was that conservation laws (upon which the entire structure of modern science is built) were not valid when applied to regions of subatomic dimensions. Another novel explanation, but one which would maintain the integrity of the conservation laws, was a proposal by Wolfgang Pauli in 1933 which hypothesized a new and fundamental particle to account for the loss of energy from the nucleus. This particle would be emitted by the nucleus simultaneously with the electron, would carry with it no electric charge, but would carry the missing energy and momentum—escaping from the laboratory equipment without detection.

Except for tenacious faith in the "conservation laws," Pauli would have had no reason to propose the existence of his ghostly particle and Fermi[11] and Reines and Cowan and the others who developed the theory and experiments which ultimately led to the experimental proof of the existence of the neutrino would have had little reason to take the proposal seriously.

Of course, some of our surest and most useful science was at one time heresy to the tenets of the savants. But it seems most unlikely that one will successfully challenge such a reliable[12] science as thermodynamics—certainly, at least, until one has learned its precepts and lived by its discipline.

[8] Frederick Reines and Clyde L. Cowan, Jr. *Nature*, **178**, 446 (1956).

[9] F. Reines and C. L. Cowan, Jr., *Phys. Rev.*, **92**, 830 (1956).

[10] W. Pauli in Rapports du Septiem Conseil de Physique Solvay, Brussels, 1933 Gautier-Villars, Paris (1934).

[11] E. Fermi, *Z. Physik* **88**, 161 (1934).

[12] See "Thermodynamics in Einstein's Thought" by M. J. Klein, *Science*, **157**, 509 (1967).

TEMPERATURE AND THE MACROSCOPIC BEHAVIOR OF GASES

Although the principle of conservation of energy does apply to microscopic and subatomic processes, the concept of heat and the enormously varied and useful relationships which can be deduced by combining this principle with the Second Law and one of two other basic postulates of thermodynamics have meaning only with respect to macroscopic systems. Classical thermodynamic theory requires no knowledge[13] of the fine structure of matter or the distribution of energy within this fine structure, and, alone, it can give no information about these things. In later sections of this book it is shown how by statistical treatments of the quantized energy states of atoms and molecules thermodynamic properties of some substances can be calculated from spectroscopic data. Yet although the theory of these calculations is a part of atomic and molecular theory, their results apply only to systems containing large numbers of molecules.

That thermodynamics is a science of the macroscopic system is evident from the fact that temperature is one of the basic thermodynamic variables. Temperature and ideas like constancy of temperature, uniformity of temperature, and equality of temperature have little meaning in relation to few-atom systems. But they have great significance with respect to larger systems. What is the meaning of the temperature of a macroscopic system?

In the preceding section energy was discussed with reference to basic experiments and their results, and a quantitative definition of energy was given in terms of heat and of various kinds of work which can be measured by comparison with the amount of some form of work required to produce the same effect. In this section a somewhat similar treatment of temperature is given. The quantitative expression that we present at this point serves our present purposes. Later, after the Second Law has been introduced, a much broader and more fundamental basis will be available for defining a scale of temperature, but it will turn out that the scale defined on that basis is identical to the one we define in this section on the basis of the behavior of gases.

Every change is driven by some kind of a potential difference. The potential which causes heat to flow (causes a thermal transfer of energy) from one system into another with resulting changes in most of the properties of one or both systems is called temperature. One of the characteristics that temperature shares with other kinds of potentials is that if *A* is in thermal

---

[13] However we find that some kinds of reasonable assumptions about the fine structure, particularly, for example, about distinctions among nonionic and various types of ionic substances, are very useful in setting up convenient thermodynamic notations. In fact we would probably view the Debye–Hückel theory as contributing considerably more than a convenient notation to the thermodynamic treatment of electrolyte solutions.

equilibrium with B and A is also in thermal equilibrium with C, B and C are in states corresponding to thermal equilibrium with each other. It is perhaps not immediately obvious why this almost-to-be-expected relationship should sometimes be dignified by the name of "Zeroth Law" of thermodynamics—if indeed that name is a dignity—but it is the useful fact underlying thermometry.

PROBLEM 1-2   An Exercise in Logic. The First Law denies the possibility of creating or destroying energy. The Second Law denies the possibility of making a perpetual motion machine which, while not involving the creation of energy, would permit an unending cyclic process of heat-to-work and work-to-heat interconversions.

Suppose, contrary to the Zeroth Law, it were possible to have A in thermal equilibrium with B and simultaneously with C but that when B and C were put into direct thermal contact they would be found not to be in thermal equilibrium with each other and heat would flow between them.

Try to devise a perpetual motion machine or one which would create or destroy energy based upon this supposition. If such a machine could be devised, then the supposed situation is ruled out by the First and Second Laws, and the Zeroth Law is a corollary of the First and Second Laws and is not an independent postulate. If the Zeroth Law could be violated *without* violating the First or Second Laws then it *is* an independent postulate and in that sense logically considered one of the few foundation principles of thermodynamics.

::

Any system having a conveniently measurable property which varies with the temperature could serve as the basis of a temperature scale. However, an especially advantageous scale is one based on the elastic properties of gases.

In 1660 Robert Boyle reported experiments indicating that at a constant temperature the pressure and volume of a confined sample of gas were related by:

$$PV = \text{constant [Boyle's Law]} \tag{1-11}$$

More than a century later Jacques Charles and Joseph Louis Gay-Lussac did experiments on the effect of changing the temperature of confined samples of different gases. Their results can be summarized by:

$$\frac{(PV)_{T_2}}{(PV)_{T_1}} = \text{constant (for different gases)} \quad \begin{bmatrix} \text{Law of Charles} \\ \text{or Gay-Lussac} \end{bmatrix} \tag{1-12}$$

Note that in the relation indicated by 1-12 it is not assumed that numerical values are known for T_1 and T_2 but only that T_1 and T_2 remain the same for the experiments on the different gases. A verbal statement of this law is that the fractional increment in PV corresponding to changing the temperature

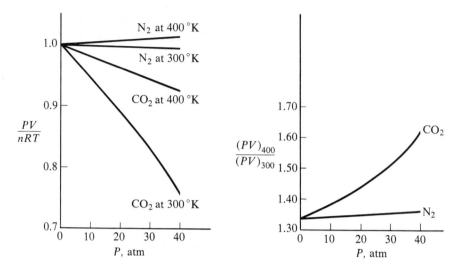

FIGURE 1-2   These graphs illustrate the approach of real gases to ideal gas behavior (Equations 1-11 and 1-12) as the pressure approaches zero.

of a gas from T_1 to T_2 is found to be the same for samples of any gases subjected to this change in temperature.

Refined experiments show that 1-11 and 1-12 are only approximately true for the ordinary gases under ordinary conditions and that deviations are large for gases at high pressure or near condensation conditions. But by extrapolating gas behavior to zero pressure, relations 1-11 and 1-12 become exactly valid. A gas which conforms to those relations is called an *ideal gas*, and any real gas approaches the behavior of ideal gas as its pressure approaches zero.

The ideal gas temperature scale is one in which the value of the temperature is arbitrarily set proportional to the value of PV for an ideal gas

$$T = \text{constant} \times PV \qquad (1\text{-}13)$$

or

$$\frac{T_2}{T_1} = \frac{(PV)_2}{(PV)_1} \qquad (1\text{-}14)$$

The only additional requirement for completing the definition of the temperature scale is explicitly or implicitly to make an arbitrary assignment of one temperature. The presently accepted assignment is that the temperature of the triple point of water is to be given the value of 273.1600 °K.

Having established this ideal-gas temperature scale one can immediately write an equation of state for the ideal gas. An equation of state is one that indicates the relation between P, V, and T for a substance. Taking account of Avogadro's law we can write

$$PV = nRT \tag{1-15}$$

where n is the number of moles of gas and R is a constant which is the same for all ideal gases and whose value can be determined (for example) by measuring PV and n at 273.16 °K.

> *PROBLEM 1-3 Avogadro's Law.* What is Avogadro's law? In what way, if any, is Equation 1-15 based on it? What is the definition of a mole? Should 1-15 be considered the definition of the value of n for an ideal gas?

THE FIRST LAW AND THE MACROSCOPIC BEHAVIOR OF GASES

Any introduction to thermodynamics must include examples of its applications. Moreover, it seems advantageous to begin exercises in thermodynamic reasoning as soon as some of the principles on which this type of reasoning is based have been introduced—not to postpone such exercises until each of these basic postulates, even though there are only a few of them, have been set forth. In fact, some of the rather simple First-Law treatments of gas systems facilitate an understanding of and lead to a quantitative expression for the Second Law. Therefore, as we continue to develop the basic structure of thermodynamics, we explore somewhat further the macroscopic behavior of gases, emphasizing, however, that the generality of thermodynamics is not limited either by our provisional use of the ideal gas to establish a temperature scale or by the use of any particular types of substances or machines which we find convenient in proving thermodynamic theorems and developing thermodynamic relations.

The distinction between a reversible and an irreversible process is important in thermodynamics. A satisfactory criterion for the reversible process is that it be possible to reverse the *direction* of the process at any moment by an *infinitesimal* change in some variable. This implies that all changes which occur in a reversible process must occur infinitesimally close to equilibrium conditions. For example cylinder-piston machines which we use in expansion problems must be free of friction if the expansion is to be reversible. Friction would prevent reversal of the direction of motion of the piston by any infinitesimal change in the pressure on either side of it. Friction would require that the pressure under the piston be *finitely* greater than the pressure over it before it would move up, and vice versa before it would move down.

In heat transfer processes which are reversible, there can exist at any moment between any parts of the system(s) or surroundings that are involved in the process only infinitesimal temperature differences. In reversible processes involving the flow of electricity there can exist only infinitesimal differences in electrical potential. We find the same kind of statement can be made in terms of the chemical potential about chemical reactions occurring reversibly.

Any processes which do not satisfy the reversibility criterion are irreversible. This does not mean that systems which undergo irreversible processes cannot be restored to their original state. The fact that they can, in many cases, makes our definition of the term *reversible process* seem rather arbitrary at present. The classification is useful, however, and later it will become clear that it is never possible to restore everything (system and surroundings) to its former condition after an irreversible process has occurred.

PROBLEM 1-4   Starting with the basic expression for the work of moving an object a distance dx against a force f show that the work associated with expansion is

$$dw = -p \, dV \tag{1-16}$$

::

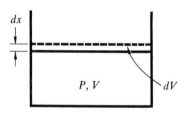

FIGURE 1-3

PROBLEM 1-5   By making use of the equation of state of an ideal gas (1-15) integrate 1-16 to show that the work associated with an isothermal reversible process whereby n moles of an ideal gas goes from $P_1 V_1 T$ to $P_2 V_2 T$ is given by the expression

$$w = nRT \ln \frac{P_2}{P_1} = nRT \ln \frac{V_1}{V_2} \tag{1-17}$$

::

PROBLEM 1-6   An ideal gas is expanded by heating at constant pressure from T_1 to T_2. Show that the work associated with this process is

$$w = nR(T_1 - T_2) \tag{1-18}$$

::

PROBLEM 1-7 A mole of liquid is evaporated at the temperature T and at the constant pressure of 1 atm. Show that the work associated with this process should be roughly

$$w = -RT \qquad (1\text{-}19)$$

State the approximations made in getting this equation.

∷

There are several kinds of processes which are of especial importance in thermodynamics. These are (1) the constant temperature, (2) the constant volume, (3) the constant pressure, (4) the adiabatic, and (5) the reversible (and the irreversible) processes. The adiabatic process is nothing more or less—is defined as—one in which no heat is transferred across the boundary of the system, i.e., one for which $q = 0$. Adiabatic processes are achieved in theory and approximated in experiment by having the system contained in a thermally insulated container. It might profitably be pointed out that adiabatic processes are very rarely isothermal; the two classifications should not be confused with each other.

The First Law equation becomes simplified for two of these cases. For the constant volume process in the common case where electrical or other forms of work other than expansion work are excluded, it becomes

$$\Delta E = q \text{ (constant volume)} \qquad (1\text{-}20)$$

because by Equation 1-16, w for this case is zero. Likewise for the adiabatic process

$$\Delta E = w \text{ (adiabatic)} \qquad (1\text{-}21)$$

Neither of these kinds of processes is the most common experimental one. The constant pressure process is, because of the fact that it is easier to operate on systems in equilibrium with the atmosphere than otherwise. It turns out that for constant pressure processes many thermodynamic equations assume a somewhat simpler form if the energy E is replaced by what may at the outset be taken as an arbitrarily defined function of E and P and V which function is given the symbol H. H is defined by the equation

$$H \equiv E + PV \qquad (1\text{-}22)$$

H has been given several names. In most cases we use the letter itself. If we use a name it will usually be *enthalpy*. Any attributes of H or equations involving it must be derived basically through defining Equation 1-22.

It was pointed out above that thermodynamics is concerned with changes in E, not its absolute value. Equation 1-22 makes it clear, then, that we are not concerned with absolute values of H and that our useful working definitions are

$$\Delta H = \Delta E + \Delta(PV) \quad \text{(finite change)} \tag{1-23}$$

and

$$dH = dE + d(PV) \quad \text{(infinitesimal change)} \tag{1-24}$$

Because E, P, and V are properties and dE, dP, and dV are exact differentials, H also is a property and dH an exact differential. H is one of the eight classical thermodynamic variables. The others are PVT and E, which we have already encountered, and SA and G which are defined later.

Initial state   Final state

FIGURE 1-4

PROBLEM 1-8 Proof that H is not conserved. A mixture of H_2 and Cl_2 is contained in a constant volume container which is thermally insulated from the surroundings so that the gas mixture can be considered an isolated system. A spark, the energy of which may be considered negligible, ignites the mixture. From a consideration of this change in state prove that, unlike the energy E, the enthalpy H does not necessarily remain constant in an isolated system.

∷

PROBLEM 1-9 ΔH in the constant pressure process. Starting with the defining equation (1-22 or 1-23) show that for any constant pressure process in which no work other than expansion work is involved

$$\Delta H = q \quad \text{(constant pressure)} \tag{1-25}$$

∷

A very fundamental property of systems is the heat capacity. This is the relation between the amount of heat that enters the system and the temperature rise that results. The symbol for heat capacity is C and the defining equation is

$$C = \frac{dq}{dT} \tag{1-26}$$

in which dq represents an infinitesimal quantity of heat entering a system and dT is the resulting temperature increment. For a given system the value of C depends on the constraints imposed on the system. The two cases of common interest are the constant volume case and the constant pressure case; the corresponding heat capacities are defined by the equations

$$C_V = \left(\frac{\partial q}{\partial T}\right)_V \tag{1-27}$$

and

$$C_P = \left(\frac{\partial q}{\partial T}\right)_P \tag{1-28}$$

PROBLEM 1-10   Show that

$$C_V = \left(\frac{\partial E}{\partial T}\right)_V \tag{1-29}$$

and

$$C_P = \left(\frac{\partial H}{\partial T}\right)_P \tag{1-30}$$

■■

Relations may readily be derived between C_P and C_V. The following is such a derivation. For the homogeneous closed system under ordinary conditions, specification of two properties such as T and V suffices to fix the state of the system. That is, we can consider E a function of two independent variables such as T and V and write

$$dE = \left(\frac{\partial E}{\partial V}\right)_T dV + \left(\frac{\partial E}{\partial T}\right)_V dT \tag{1-31}$$

But

$$dE = dq + dw \tag{1-10}$$

For the kind of experiment relevant to the concept of heat capacity

$$dw = -P\,dV \tag{1-16}$$

So, rearranging, we can write

$$dq = dE + P\,dV \tag{1-32}$$

Combining 1-31 and 1-32 we have

$$dq = \left[\left(\frac{\partial E}{\partial V}\right)_T + P\right] dV + \left(\frac{\partial E}{\partial T}\right)_V dT \tag{1-33}$$

If we divide 1-33 by dT and indicate that P is held constant we get

$$\left(\frac{\partial q}{\partial T}\right)_P = \left[\left(\frac{\partial E}{\partial V}\right)_T + P\right]\left(\frac{\partial v}{\partial T}\right)_P + \left(\frac{\partial E}{\partial T}\right)_V \tag{1-34}$$

Now making use of 1-28 and 1-29 and rearranging, we have

$$C_P - C_V = \left[\left(\frac{\partial E}{\partial V}\right)_T + P\right]\left(\frac{\partial V}{\partial T}\right)_P \tag{1-35}$$

PROBLEM 1-11   By a similar procedure derive the result

$$C_P - C_V = \left[V - \left(\frac{\partial H}{\partial P} \right)_T \right] \left(\frac{\partial P}{\partial T} \right)_V \tag{1-36}$$

The partial derivatives in 1-35 and 1-36 are properties of substances which have ultimately to be determined experimentally. We already know something about the behavior of gases and for the ideal gas we can readily evaluate the $(\partial V/\partial T)_P$ of 1-35 from the equation of state (1-15). We next consider an experiment of James Joule on the basis of which we can evaluate the property $(\partial E/\partial V)_T$ for an ideal gas.

His experiment was to let compressed air expand into a vacuum and to measure the net heat effect. The experiment and its results are shown schematically in Figure 1-5. The water surrounding the expansion apparatus

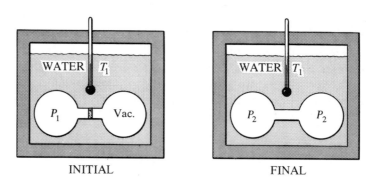

INITIAL   FINAL

FIGURE 1-5

was contained in a thermally insulated vessel. Cooling would occur in the bulb in which the pressure decreased and heating in the other, but when thermal equilibrium had been reattained, the final temperature of the bath was found to be the same as its initial temperature. Thus $q = 0$, $\Delta T = 0$ and also, since no work has passed between the gas and its surroundings, $w = 0$. Since q and w are both zero, $\Delta E = 0$.

It is difficult to do this experiment with great precision because the heat capacity of the bath is large compared to any heat effect that might be found. Related kinds of experiments (the Joule–Thomson experiment, for example) which can be done with greater precision show that the above-stated result is characteristic exactly only of gases at pressures approaching zero, i.e., it is characteristic of the ideal gas. The property shown to be zero for the ideal gas in this experiment may be described by the partial derivative $(\partial T/\partial V)_E$ and we can write

$$\left(\frac{\partial T}{\partial V} \right)_E = 0 \quad \text{(for ideal gas)} \tag{1-37}$$

PROBLEM 1-12 By dividing 1-31 through by dV and considering the case in which E is constant show that 1-37 leads to the conclusion that for an ideal gas

$$\left(\frac{\partial E}{\partial V}\right)_T = 0 \quad \text{(ideal gas)} \tag{1-38}$$

::

PROBLEM 1-13 Use 1-29 and 1-38 to show that for an ideal gas C_V may be a function of T but at a constant temperature is independent of volume.

::

PROBLEM 1-14 From 1-35 and other available relationships show that for an ideal gas, per mole,

$$C_P - C_V = R \quad \text{(ideal gas)} \tag{1-39}$$

::

PROBLEM 1-15 Justify the statement that for an ideal gas, whether or not the volume is held constant,

$$dE = C_V \, dT \quad \left(\begin{array}{l}\text{For ideal gas}\\ \textit{whether or not}\\ \text{volume is}\\ \text{constant}\end{array}\right) \tag{1-40}$$

::

The Joule experiment is an irreversible process. In theory it would be possible to have the process occur adiabatically by letting heat redistribute itself after the expansion by conduction and convection within the gas itself and without using the water bath to transfer it. We would conclude that in *that kind* of irreversible adiabatic expansion of an ideal gas, $T_2 - T_1 = 0$. (It should be pointed out that there are many—in fact, with apologies for the sensational sound of the term, an infinite number of—different irreversible processes of carrying out a given change in state.)

Let us now derive a relation between T_1, T_2, V_1 and V_2 for the *reversible* adiabatic expansion of an ideal gas.

$$dE = dw \quad \text{[Adiabatic]} \tag{1-21}$$

$$dE = C_V \, dT \quad \left[\begin{array}{l}\text{Ideal gas whether or}\\ \text{not } V \text{ is constant}\end{array}\right] \tag{1-40}$$

$$dw = -P \, dV \quad \text{[No electrical work]} \tag{1-16}$$

$$dw = -\frac{nRT}{V} \, dV \quad \text{[Reversible, ideal gas]} \tag{1-41}$$

By combining the above equations and indicating integration over a finite range of T and V we get

$$\int_{T_1}^{T_2} C_V \frac{dT}{T} = -nR \int_{V_1}^{V_2} \frac{dV}{V} \tag{1-42}$$

In order to carry out the integration, the approximation is usually made that C_V is a constant over the required temperature range. With this approximation the result is

$$C_V \ln \frac{T_2}{T_1} = nR \ln \frac{V_1}{V_2} \quad \begin{pmatrix} \text{Reversible} \\ \text{adiabatic} \\ \text{expansion} \\ \text{of ideal} \\ \text{gas} \end{pmatrix} \tag{1-43}$$

PROBLEM 1-16   A commonly given expression for the relations in a reversible adiabatic expansion of an ideal gas is

$$PV^\gamma = \text{constant} \quad \begin{pmatrix} \text{Reversible adiabatic} \\ \text{expansion of ideal} \\ \text{gas} \end{pmatrix} \tag{1-44}$$

where $\gamma = C_P/C_V$. Derive 1-44 from 1-43 and other necessary relations.

::

One has a large choice of equations which express approximately the *PVT* relations of gases at moderate pressures. Our interest in either ideal or nonideal gases is with respect to their use in thermodynamic reasoning. As equations of state of nonideal gases at moderate pressures we usually choose van der Waals' equation and equations that can be derived from it. His equation is, per mole

$$\left(P + \frac{a}{V^2}\right)(V - b) = RT \quad \begin{pmatrix} \text{van der Waals'} \\ \text{equation} \end{pmatrix} \tag{1-45}$$

PROBLEM 1-17   By solving for P from 1-45 and expanding the fraction $1/(V - b)$ into a power series in $1/V$ show that van der Waals' equation is equivalent to the virial equation (power series)

$$PV = RT\left(1 + \frac{A}{V} + \frac{B}{V^2} + \cdots\right) \tag{1-46}$$

::

PROBLEM 1-18   From 1-46 show that an approximate form of van der Waals' equation is

$$PV = RT + \alpha P \tag{1-47}$$

::

PROBLEM 1-19   By considering the solution to Problems 1-17 and 1-18 derive an expression for α in terms of the constants in van der Waals' equation and the temperature

Answer:
$$\alpha = b - \frac{a}{RT} \tag{1-48}$$

::

PROBLEM 1-20 (a) Derive an expression in terms of T, V_1, V_2, and α for the work of expanding reversibly and isothermally from V_1 to V_2 a mole of a gas which obeys 1-47.

(b) For the same case get an expression for w in terms of T, P_1, P_2, and α.

▪▪

The Joule–Thomson experiment is represented in Figure 1-6. It is sometimes called the porous-plug experiment because two piston-cylinder apparatus are connected through a porous plug or other connection which impedes the flow of gas. The pistons and cylinders may actually be far apart and one of the pistons might be just the atmosphere. But for

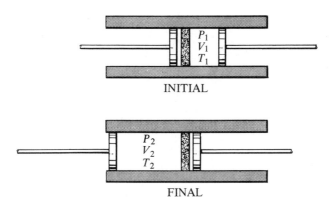

INITIAL

FINAL

FIGURE 1-6

analyzing the experiment the scheme shown is convenient. The experiment is carried out adiabatically. At the start, all of the gas is on the right side of the porous plug. The piston to the right is pushed toward the plug and other piston, initially against the plug, is pulled away from it. The rates of motion of the pistons are such as to keep the pressure on the right side constant at P_1 and on the left at some lower constant pressure P_2. We consider the experiment terminated when the piston on the right side has come up to the plug.

The experimental observation is the temperature change, if any, which the gas undergoes in passing through the plug. The finding is that there are, below certain limiting pressures, two "inversion temperatures," characteristic of each gas and dependent upon the pressure. Between the upper and lower inversion temperatures dT/dP is negative; below the lower and above the upper inversion temperature, dT/dP is positive; at the inversion temperatures $dT/dP = 0$.

PROBLEM 1-21   The Joule–Thomson coefficient   Starting with the definition of H and by considering the work associated with the Joule–Thomson experiment (see Fig. 1-5), show that in this experiment, regardless of the properties of the fluid forced through the porous plug, $\Delta H = 0$. Because of this result, the Joule–Thomson coefficient μ_{JT} can be defined by

$$\mu_{JT} = \left(\frac{\partial T}{\partial P}\right)_H \tag{1-49}$$

■■

PROBLEM 1-22   Manipulation of partial derivatives   In ordinary cases including the Joule–Thomson experiment, H may be considered a function of any pair of independent properties, T and P, for example. By writing

$$dH = \left(\frac{\partial H}{\partial P}\right)_P dT + \left(\frac{\partial H}{\partial P}\right)_T dP \tag{1-50}$$

and setting $dH = 0$, show that

$$\left(\frac{\partial T}{\partial P}\right)_H = -\frac{\left(\frac{\partial H}{\partial P}\right)_T}{\left(\frac{\partial H}{\partial T}\right)_P} \tag{1-51}$$

and hence that

$$\mu_{JT} = -\frac{1}{C_P}\left(\frac{\partial H}{\partial P}\right)_T \tag{1-52}$$

■■

THERMOCHEMISTRY

It has been pointed out that for a given change in state, q and w may have different values depending upon the process of effecting the change. However, under certain restrictions the total heat effect accompanying a specified over-all chemical reaction will be the same regardless of possible variations of the intermediate steps in the reaction. The two cases of usual interest are of the reaction carried out at constant volume and of the reaction carried out at constant pressure.

For the constant volume case, $w = 0$ and $q = \Delta E$. Inasmuch as ΔE must remain constant for any possible variations in the intermediate steps of a chemical reaction the summation of the q's over any possible series of intermediate steps must also be constant. The constant pressure case is almost as straightforward. The net work must be $-P\,\Delta V$ in all cases. Because $q = \Delta E + P\,\Delta V$, again the sum of the q's must be constant for all possible paths of achieving a given reaction. These facts are sometimes called Hess' law of the additivity of heats of reaction.

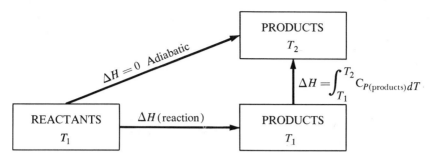

FIGURE 1-7

Heats of reaction, whether in terms of energies (ΔE's) or enthalpies (ΔH's), always refer to an isothermal reaction, i.e., one in which the initial state is of reactants at T and the final state is of the products at T. However, the measurements of these heats is often done by "adiabatic calorimetry." In adiabatic calorimetry the observation is of the temperature rise of the system which accompanies the reaction carried out adiabatically. The relationships are easily seen from Figure 1-7 for the constant pressure case. The relation is

$$-\Delta H_{(\text{reaction at } T_1)} = \int_{T_1}^{T_2} C_{P(\text{products})} \, dT$$

An entirely analogous result holds in terms of ΔE and C_V for the constant volume case. It should be pointed out that what we have denoted as $C_{P(\text{products})}$ actually includes the heat capacity of the calorimeter and everything else that is present in the system after the reaction has occurred. It should also be observed that the integral in the above equation need not be evaluated from an explicit knowledge of C_P. One can determine the electrical energy required in an adiabatic process to raise the temperature of the "products" from T_1 to T_2 and this, by the First Law, must be exactly equivalent to the heat required in the thermal (nonadiabatic) process.

The orderly and convenient way of making thermochemical calculations is based on the concept of the standard enthalpy of formation, ΔH_f°.[14] The definition of the standard enthalpy of formation of a compound is indicated in Figure 1-8. The standard state is usually taken as the stable

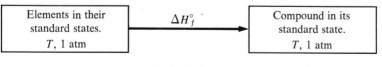

FIGURE 1-8

[14] From here on, it should be unnecessary to point out in cases such as this the parallel considerations that apply to ΔE.

form of the element or compound at the temperature of interest, T, and under 1 atm pressure. However, standard states are whatever people choose to make them and sometimes it is convenient to use other than the above stated convention.[15] The standard state referred to should be specifically indicated, especially if it is not the conventional one.

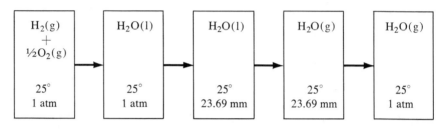

FIGURE 1-9

PROBLEM 1-23 The conventional standard state for water at 25° is liquid water at 1 atm. The value at 25 °C for ΔH_f for $H_2O(l)$ is -68.3174 kcal mole^{-1}. The vapor pressure of water at 25° is 23.69 mm and its heat of vaporization at the temperature is 582.3 cal g^{-1}. By considering the steps indicated in Figure 1-9, calculate the value for ΔH_f for $H_2O(g)$ at 25°, 1 atm. This is not an experimentally attainable state of water, but the value of ΔH_f° (-57.7979 cal mole^{-1}) for this hypothetical state may be used conveniently and correctly in practical calculation (see Problem 1-26).

::

The utility of standard enthalpies of formation in calculations about heats of reaction is indicated in Problem 1-24.

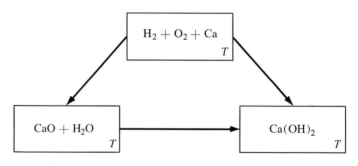

FIGURE 1-10

[15] In dealing with nonideal gases and with solutes in liquid solutions a different choice of standard states is conventional. Additional consideration is given to definitions of standard states in following sections of this book.

PROBLEM 1-24 By considering the definition of ΔH_f° and the change-of-state diagram in Figure 1-10, show that for the reaction

$$CaO + H_2O \rightarrow Ca(OH)_2$$

$$\Delta H_{reaction} = \Delta H_f^\circ(Ca(OH)_2) - \Delta H_f^\circ(CaO) - \Delta H_f^\circ(H_2O)$$

and that in general,

$$\Delta H_{(reaction)}^\circ = \Delta H_{f(products)}^\circ - \Delta H_{f(reactants)}^\circ \qquad (1\text{-}53)$$

▪▪

Table 1-1   Standard Enthalpies of Formation at 25 °C*

Values of ΔH are in kcal mole^{-1}

Compound	ΔH_f°	Compound	ΔH_f°	Compound	ΔH_f
AgCl(c)	−30.362	CaO(c)	−151.9	H$_2$O(l)	−68.317
Ag$_2$O(c)	−7.306	CuO(c)	−37.1	HgO(c)	−21.68
CH$_3$OH(l)	−57.036	CuSO$_4$(c)	−184.00	Hg$_2$Cl$_2$	−63.32
CH$_4$(g)	−17.889	CuSO$_4$·5H$_2$O(c)	−544.45	NH$_3$(aq)	−19.32
C$_2$H$_2$(g)	54.194	HBr(g)	−8.66	NH$_3$(g)	−11.04
C$_6$H$_6$(l)	11.718	HCl(aq)	−40.023	NO(g)	21.600
CO(g)	−26.4157	HCl(g)	−22.063	NO(g)	8.091
CO$_2$(g)	−94.0518	HCN(g)	31.2	NaCl(c)	−98.232
CaCO$_3$(c)	−288.45	HI(g)	6.2	TlCl(c)	−48.99
(Calcite)					
CaCO$_3$(c)	−288.49	H$_2$O(g)	−57.798	ZnO(c)	−83.17
(Aragonite)					

* From Rossini, Wagman, Evans, Levine and Jaffe, "Selected Values of Chemical Thermodynamic Properties," National Bureau of Standards, Circular 500, 1949.

The dependence of enthalpies of reaction upon the temperature at which the reaction is carried out may be seen from Figure 1-11.

$$\Delta H_{T_2} = \Delta H_{T_1} - \Delta H_R + \Delta H_P$$

$$\Delta H_R = \int_{T_1}^{T_2} C_{P(reactants)}\, dT$$

$$\Delta H_P = \int_{T_1}^{T_2} C_{P(products)}\, dT$$

Hence

$$\Delta H_{T_2} - \Delta H_{T_1} = \int_{T_1}^{T_2} \Delta C_P\, dT \qquad (1\text{-}54)$$

In order to get the relation between enthalpies of reaction at two finitely different temperatures, 1-54 must be integrated. In some cases a

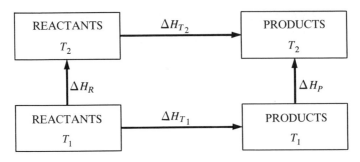

FIGURE 1-11

satisfactory approximation is that ΔC_P is a constant over the range of temperature between T_1 and T_2. In other cases experimental heat capacities may be expressible to good approximation over a considerable temperature range as a standard function of the temperature which contains several constants whose values are adjusted to fit the heat capacity—temperature data of the substance in question. For example, in Glassner's[16] tabulation of the thermochemical properties of the oxides, fluorides, and chlorides to 2500 °K an equation of the form

$$C_P = a + bT + cT^2 + \frac{d}{T^2} \tag{1-55}$$

is used. For a reaction for which the C_P's of all the reactants and products are satisfactorily expressed by 1-55 one could write an analogous expression for ΔC_P, namely,

$$\Delta C_P = \Delta a + \Delta bT + \Delta cT^2 + \Delta d\frac{1}{T^2} \tag{1-56}$$

PROBLEM 1-25   On the basis of 1-54, 1-55, and 1-56 show that

$$\Delta H_T = \Delta H_{298} + \Delta aT + \tfrac{1}{2}\Delta bT^2 + \tfrac{1}{3}\Delta cT^3 - \frac{\Delta d}{T} + \Delta A \tag{1-57}$$

where ΔA is a constant.

■■

PROBLEM 1-26   The maximum temperature which is theoretically possible to obtain from an exothermic chemical reaction would be that temperature which would correspond to carrying the reaction out under adiabatic conditions (no heat given to surroundings).

Figure 1-7 serves to indicate the relationships. In fact, the calculation of maximum (adiabatic) flame temperature is almost exactly

[16] Alvin Glassner, U.S. Atomic Energy Commission Bulletin ANL 5750, (1957).

analogous to the calculations in adiabatic calorimetry. These latter calculations are usually simplified by the fact that temperature ranges are small enough so that heat capacities may be satisfactorily treated as constants. This is not so in the calculation of flame temperatures.

Taking the following values[17] for a, b, c, d of 1-55 for $H_2O(g)$ and N_2,

	a	$b \times 10^3$	$c \times 10^6$	$d \times 10^{-5}$
H_2O	7.30	2.46	—	—
N_2	6.76	0.606	0.13	—

and using the answer to Problem 1-23, estimate the maximum temperature that could result from burning hydrogen in air. Note that $C_{P(products)}$ here must include the heat capacity of the unreacting nitrogen which goes along for the ride.

::

AN IDEAL GAS UNDERGOING A CARNOT CYCLE

As our last exercise before considering the Second Law, we describe the Carnot cycle and derive some relations for the case that an ideal gas is put through this cycle. The cycle is illustrated by Figure 1-12. The essential feature of a Carnot cycle is that heat is exchanged between system and surroundings at only two temperatures, T_1 and T_2, and that the changes in temperature of the system from T_1 to T_2 and then from T_2 to T_1 are accomplished adiabatically.[18]

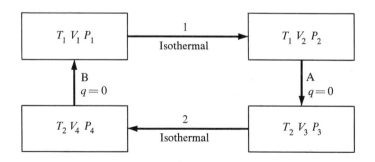

FIGURE 1-12   The Carnot Cycle.

[17] These values are such that C_P comes out in cal deg^{-1} mole^{-1}.

[18] There is in general no requirement that the P's and V's should vary as indicated in Figure 1-12. For example, it would be possible to operate an electrical system comprising a motor, a generator and a storage cell in a Carnot cycle all at constant pressure. At least we use the term Carnot cycle with this general significance and do not restrict it to expansion engines.

Suppose that an ideal gas is expanded and compressed *reversibly* in a Carnot cycle. The values of q and w for each step can be evaluated and the expression for the interesting quantity w/q_1 can be derived. Here w is the total work around the cycle and w/q_1 is called the efficiency of the Carnot-cycle machine.[19] The importance of the result of this derivation becomes clear not only when, later, we state a basis for concluding that exactly the same result (the expression for w/q_1) would apply to a steam engine or to the electrical system suggested in footnote 18, but even more so when it is shown that this result can be used as the chief link between very simple statements of the First and Second Laws and a fascinating variety of quantitative conclusions concerning physical and chemical equilibria.

Consider an ideal gas put reversibly through a Carnot cycle. Starting at T_1V_1 and P_1 it is first (in step 1) expanded isothermally to V_2 and P_2. In Step A it is further expanded, but adiabatically, until it has the properties T_2, V_3, and P_3. The next step (step 2) is an isothermal compression at T_2, and the final step (step B) is an adiabatic compression to restore the gas to its initial properties.

PROBLEM 1-27 By considering the properties of an ideal gas, argue that $w_1 = -q_1$ and $w_2 = -q_2$ (review 1-38 if necessary). Show also that $w_A + w_B = 0$ (review 1-21, 1-40 and Problem 1-12 if necessary). Thus show that

$$\frac{-w}{q_1} = \frac{-(w_1 + w_2)}{q_1} = \frac{T_1 \ln \dfrac{V_2}{V_1} + T_2 \ln \dfrac{V_4}{V_3}}{T_1 \ln \dfrac{V_2}{V_1}} \tag{1-58}$$

::

PROBLEM 1-28 By noting as shown in Problem 1-27 that $w_A + w_B = 0$ and that each of these w's can be written as $-w = nR \int T(dV/V)$ show that

$$\ln \frac{V_3}{V_2} = -\ln \frac{V_1}{V}$$

(review 1-42 if necessary) and hence that

$$\frac{-w}{q_1} = \frac{T_1 - T_2}{T_1} \quad \begin{pmatrix} \text{Efficiency of} \\ \text{a reversible} \\ \text{Carnot cycle} \end{pmatrix} \tag{1-59}$$

::

[19] The overall efficiency of any cyclic device whether operating reversibly or irreversibly must be unity as required by the First Law. That is, around any cycle the net work done on the system plus the net heat put into it must be zero and the ratio of these quantities, therefore, must be -1. The efficiency ratio w/q_1 or $-w/q_1$ on the other hand is a measure of relative magnitudes of the work done on or by the system to the heat put into it at T_1, and without explicit representation of the fact that the energy deficit is made up by a heat flow q_2 at T_2. These ratios may be considered expressions for the effectiveness of the machine in transferring heat by the use of work or of doing work by the use of heat.

PROBLEM 1-29   Show that for a reversible Carnot cycle, 1-27 leads to the conclusion that (since $-w = q_1 + q_2$)

$$\frac{q_1}{T_1} + \frac{q_2}{T_2} = 0 \tag{1-60}$$

■■

As was mentioned earlier, one can conceive of a variety of engines in addition to the ideal gas one which could be operated in a reversible Carnot cycle. It is instructive to consider the possibility of coupling an ideal-gas Carnot engine with another Carnot engine such as a steam engine. They can be coupled in such a way that the ideal-gas engine absorbs heat from the surroundings at the temperature at which the steam engine gives heat to the surroundings, and vice versa.

FIGURE 1-13

We have already developed an expression for w/q_1 for the reversible, ideal-gas Carnot cycle. Suppose that the factor relating the efficiency (w/q_1) of the ideal gas engine to the steam engine is k. Then for the steam engine

$$-\frac{w}{q_1} = k\left[\frac{T_1 - T_2}{T_1}\right] \quad \begin{bmatrix} \text{For hypothetical} \\ \text{steam engine} \end{bmatrix} \tag{1-61}$$

This can be rearranged to

$$q_1 = \frac{wT_1}{k(T_2 - T_1)} \quad \begin{bmatrix} \text{For hypothetical} \\ \text{steam engine} \end{bmatrix} \tag{1-62}$$

The same expression serves for the ideal gas engine if $k = 1$. Now suppose that the two engines are so adjusted that

$$w_{(gas)} + w_{(steam)} = 0$$

Then

$$q_{1(total)} = q_{1(gas)} + q_{1(steam)} = \left(\frac{w_{gas}T_1}{T_2 - T_1}\right)\left(1 - \frac{1}{k}\right) \tag{1-63}$$

PROBLEM 1-30   Remembering that $w_{(gas)}$ can be either positive or negative depending upon which way the reversibly operating device shown in Figure 1-13 is run, and also that if $w_{(gas)} + w_{(steam)} = 0$ for operating in one direction it will also $= 0$ for operation in the other direction, show that unless $k = 1$ the device of Figure 1-13 can be used as an instrument to transfer heat either from T_1 to T_2 or from T_2 to T_1 without an exchange of work between this composite system and its surroundings.

::

PROBLEM 1-31 By adjusting the device of Figure 1-13 so that the heat absorbed by one engine at T_2 is exactly equal to that given up by the other at T_2, show that unless $k = 1$ the device could constantly produce work from heat at T_1 without the need for any other temperature reservoir and without any noncyclic changes in the system.

::

ADDITIONAL PROBLEMS

PROBLEM 1-32 Show that if $\mu_{JT} \rightarrow$ constant (not zero) as $P \rightarrow 0$

$$\left(\frac{\partial E}{\partial V}\right)_T = 0$$

::

PROBLEM 1-33 Bearing in mind that for a van der Waals' gas at constant T, P is not inversely proportional to V show that, nevertheless, the virial equation $PV = RT(A + BP + CP^2 + \cdots)$ can be derived from van der Waals' equation. Here V is the molal volume and A, B, C, ... are constants at a constant temperature but may vary with T.

::

PROBLEM 1-34 (a) C_P for an ideal gas is given by

$$C_P = a + bT$$

Derive a relationship between T_1, T_2, V_1, and V_2 for the reversible adiabatic expansion of this gas.

(b) On the basis of the result derived in a calculate the temperature resulting from the reversible adiabatic compression of nitrogen from a volume of 10 liters to a volume of 1 liter, the initial temperature being 25 °C. Take for N_2 $a = 6.524$ cal deg^{-1} mole^{-1}, $b = 1.25 \times 10^{-3}$ cal deg^{-2} mole^{-1}.

(c) Compare the answer to b with the result of a calculation in which the heat capacity of N_2 is treated as constant at its 25 °C value.

Answer: (a) $R \ln \dfrac{V_1}{V_2} = (a - R) \ln \dfrac{T_2}{T_1} + b(T_2 - T_1)$

(b) 350 °C

::

PROBLEM 1-35(a) A tall cylinder of height h stands vertically in the earth's atmosphere. For this problem the atmosphere is considered isothermal, and it is considered a good approximation to take the acceleration of gravity as constant. P_0 is the atmospheric pressure at the *top* of the cylinder. By noting that the pressure at any distance x below the top is equal to the pressure at the top plus the pressure from the weight of the air in a column x high, derive an expression for the pressure at the point x.

Answer: $P_x = P_0 e^y$ where $y = Mx/RT$ M is the molecular weight of air. If x is in centimeters R must be in g cm mole^{-1} deg^{-1}. Ideal gas behavior is assumed in all parts of Problem 1-37.

::

PROBLEM 1-35(b) A piston is put at the top of the cylinder and the air is isothermally compressed until the piston has gone down a distance l. Show that the pressure against the piston P', is then

$$P' = P_0 \frac{(e^z - 1)}{(e^{z'} - 1)}$$

where

$$z = \frac{Mh}{RT} \quad \text{and} \quad z' = \frac{M(h - l)}{RT}$$

::

PROBLEM 1-35(c) If the piston and cylinder are 1 square centimeter in cross section, calculate the work to compress the gas to half its initial volume.

Answer: $w = -P_0(e^z - 1)\left[\frac{1}{2}h + \frac{RT}{M} \ln \frac{(e^{z/2} - 1)}{(e^z - 1)}\right]$

where P is in g cm^{-2}, w is in g cm, h is in cm, and R is in g cm deg^{-1} mole^{-1}.

::

PROBLEM 1-35(d) If $z \ll 1$ show, by expanding e^z into a convergent series, that for the 2 to 1 compression

$$w \simeq nRT \ln 2$$

::

PROBLEM 1-35(e) Use the answer to Problem 1-35(a) to calculate atmospheric pressure at an altitude of 2 km in an isothermal atmosphere at 25 °C.

::

PROBLEM 1-35(f) The cylinder of Problem 1-35(c) is filled with air and its bottom is at sea level. Calculate a numerical value for the work to carry out a 2 to 1 compression.

Answer: 11.5 kjoules

::

PROBLEM 1-35(g) Compare the answer to Problem 1-35(*f*) with the numerical value of $P_0 V_0 \ln 2$ which is the work to do a 2 to 1 compression on an ideal gas under conditions under which the effect of the gravitational field can be neglected.

::

PROBLEM 1-36 Calculate the heat evolved when 1 mole $CuSO_4(c)$ reacts with 5 moles $H_2O(l)$.

::

PROBLEM 1-37 Kelley and co-workers ("High-Temperature Heat Content, Heat Capacity and Entropy Data for the Elements and Inorganic Compounds"; Bureau of Mines Bulletin 584 (1960)) give the following empirical expressions for the constant-pressure heat capacities of O_2, CO, and CO_2 (cal mole^{-1} deg^{-1}):

$$O_2(g) \quad C_P = 7.16 + 1.00 \times 10^{-3}T - 0.40 \times 10^5 T^{-2}$$

$$CO(g) \quad C_P = 6.79 + 0.98 \times 10^{-3}T - 0.11 \times 10^5 T^{-2}$$

$$CO_2(g) \quad C_P = 10.57 + 2.10 \times 10^{-3}T - 2.06 \times 10^5 T^{-2}$$

Use these data together with data from Table 1-1 to calculate the heat of combustion of CO with O_2 at 1000 °K.

::

PROBLEM 1-38 One mole of liquid benzene is burned with oxygen to $CO_2(g)$ and $H_2O(l)$. Calculate the heat evolved, referred to 25 °C, if the reactants and products are at 1 atm (standard states).

::

PROBLEM 1-39 For the reaction

$$NO + \tfrac{1}{2}O_2 \rightarrow NO_2$$

(*a*) Calculate $\Delta H°$; (use Table 1-1)
(*b*) Calculate $\Delta E°$.

::

PROBLEM 1-40 Show that for a reaction among ideal gases ΔH and ΔE are independent of any isothermal pressure or volume changes. Show, therefore, that for such reactions, $\Delta H°$ is the heat absorbed when the reaction is carried out at constant pressure and $\Delta E°$ is the heat absorbed when the reaction is carried out at constant volume.

::

PROBLEM 1-41 Show that for a reaction among ideal gases

$$\Delta H° = \Delta E° + RT \Delta n$$

where Δn is the increase in number of moles of gas ($\Delta n = -\tfrac{1}{2}$ for the reaction given in Problem 1-39). Show also that this relation between $\Delta H°$ and $\Delta E°$

is a good approximation for reactions involving gases and liquids or solids if Δn is interpreted to mean the increase in number of moles of gas. Calculate $\Delta E°$ for the reaction

$$2Cu + O_2 \rightarrow 2CuO \qquad \text{(referred to 25 °C)}$$

::

PROBLEM 1-42   Use data from Table 1-1 to calculate the heat of vaporization of water at 25 °C.
Answer: 584 cal g^{-1}. What approximation is implicit in this calculation?
Answer: That water vapor is an ideal gas. The vapor pressure of H_2O at 25 °C is 23.8 torr. On the other hand the standard state of water vapor at 25 °C is the hypothetical 1 atm state. If the gas were not ideal one could not assume $\Delta H = 0$ for the change $H_2O(g)$ at 1 atm $\rightarrow H_2O(g)$ at 23.8 torr.

2

The Second Law

CARNOT'S PRINCIPLE

In Problems 1-30 and 1-31 it has been brought out that if there were two different kinds of machines both operable reversibly in Carnot cycles between the same two temperatures but having different efficiencies, some remarkable things could be accomplished. Problem 1-30 shows that, by means of the kind of Carnot machine proposed in Figure 1-13, if $k \neq 1$ it would be possible to transfer heat from a higher temperature reservoir to a lower temperature one *or* to transfer it in the opposite direction without any net change in the machine nor the need for any kind of energy to be put into or taken out of the machine except the heat transferred.

Problem 1-31 shows that a combination of two reversibly operating Carnot-cycle machines of different efficiencies would make possible the doing of work by a device requiring only a single constant-temperature reservoir as its source of energy. How remarkable—and useful—such a device would be has often been illustrated by pointing out that a ship using one as its motor could ceaselessly ply the seven seas, never having to stop for fuel. Such a device has been called a *perpetual motion machine of the second kind*. A perpetual motion machine of the first kind is one which would create energy. This, of course, is ruled out by the First Law.

However, all experience indicates that neither of these feats—the transfer of heat from a lower temperature to a higher one without other net changes, or the production of work from heat taken from a heat reservoir at constant temperature (also without other changes)—can be accomplished.

The First Law, as we have seen, is based not upon logical argument from some more basic postulate. Rather it represents the conclusion by scientists that the conservation of energy proven in many, many experiments and not disproven by one must be a law of nature.

The Second Law has exactly the same basis. There are many ways in which it can be correctly stated. One need start with only any one of these statements and all the others can be shown to be unavoidable consequences of it. We are now in a position to give two of these expressions. They are the conclusions that neither of the remarkable things we have speculated about in the preceding paragraphs can be accomplished, that these impossibilities represent a law of nature, the Second Law of thermodynamics.

TWO SATISFACTORY STATEMENTS OF THE SECOND LAW

1. It is impossible to bring about a net transfer of heat from a cooler to a warmer body without the simultaneous occurrence of some other net change.

2. It is impossible to make a perpetual motion machine of the second kind.[1]

In 1824 when Carnot published his "Reflexions sur la puissance motrice du feu"[2] thermodynamic theory did not exist and the Second Law had not been proposed. Although credit for the successes of any theory must properly be given not only to the one who first asserts it but also to others who bring it to light and to acceptance and who develop its consequences, classical thermodynamics owes a tremendous debt to Sadi Carnot.

It was he who first studied the cycle which bears his name. He recognized, as we have, what remarkable results could be achieved if one connected together two reversibly-operable Carnot-cycle engines which had different efficiencies. Then he wrote, "Such a creation is entirely contrary to ideas now accepted, to the laws of mechanics and of sound physics. It is inadmissible. We should then conclude that the maximum of motive power resulting from the employment of steam is also the maximum of motive power realizable by any means whatsoever."

This is Carnot's principle. Carnot's principle may be considered a statement of the Second Law.

In Problems 1-27, 1-28, and 1-29 we have already derived two quantitative relations that apply to an ideal gas undergoing a Carnot cycle reversibly. By Carnot's principle they must apply to any system. We can

[1] These two statements are equivalent to what have been called, for historical reasons, the principle of Clausius and the principle of Thomson (Kelvin).
[2] An English translation of this treatise together with translations of long papers by Clapeyron and by Clausius are available in a paperback edition published by Dover Publications, Inc., New York.

consider them as limited quantitative expressions of the Second Law for the reversible case:

$$\frac{-w}{q_1} = \frac{T_1 - T_2}{T_1} \tag{1-59}$$

and

$$\frac{q_1}{T_1} + \frac{q_2}{T_2} = 0 \tag{1-60}$$

CARNOT CYCLES WITH HEAT EXCHANGE AT MORE THAN TWO TEMPERATURES

Equations 1-59 and 1-60 are limited to cycles in which heat is exchanged with surroundings at only two temperatures. This limitation can be removed by combining cycles which operate between different pairs of temperatures. The procedure is probably best explained by graphical representation, but the question arises as to how to plot a generalized Carnot cycle. For an expansion engine one might plot isothermal and adiabatic lines on a *P–V* graph. However, it is desirable to have a more versatile plot, one which would apply to systems, for example, such as the electrochemical one we study later in this chapter, in which the pressure is constant and any volume changes are small and of little significance.

One's first thought might be to plot the total heat absorbed by the system against the temperature. This would be unsatisfactory, as shown in Figure 2-1, because the value of $\sum q_{rev}$ at the end of each cycle would be different from its value at the end of the preceding cycle. The magnitude of the difference in $\sum q_{rev}$ between successive cycles would be equal to the value

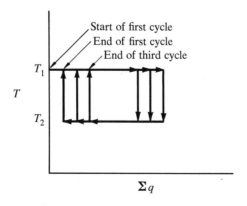

FIGURE 2-1

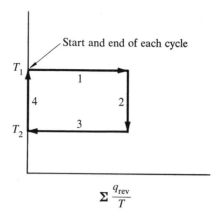

$$\Sigma \, \frac{q_{\text{rev}}}{T}$$

FIGURE 2-2   Representation of a Carnot cycle involving heat exchange at only two temperatures.

of $-w$ for the cycle, and the existence of the difference should emphasize the fact that q is not a function of state (property) of the system.

But on the other hand, if we plot $\sum (q_{\text{rev}}/T)$ against T, Equation 1-60 requires that we end at the starting point. This is satisfactory; the result is shown in Figure 2-2.

More elaborate cycles can be built up from combinations of simple Carnot cycles as shown in Figures 2-3, 2-4, and 2-5. In Figure 2-3 we have combined three cycles and have heat exchange at three temperatures.

PROBLEM 2-1   Show that a cycle around the external boundaries of the graph in Figure 2-3 is equivalent to the completion of each of the three

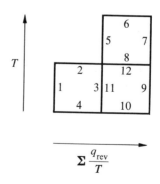

$$\Sigma \, \frac{q_{\text{rev}}}{T}$$

FIGURE 2-3.   A combination of reversible Carnot cycles with three temperatures at which heat is exchanged.

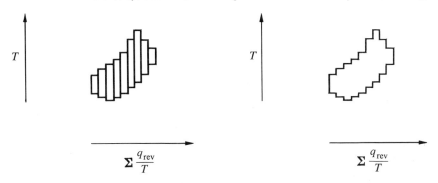

FIGURE 2-4 A composite Carnot cycle involving heat exchange at many temperatures. The graph on the left shows each component-Carnot cycle complete. The graph on the right shows only the parts of the cycles which do not cancel out.

cycles represented by the three squares. That is, show that the effects accompanying steps 3, 11, 12, and 8 cancel each other.

In Figures 2-4 and 2-5 we show how any reversible cycle can be built up from a large number, and in the limit an infinite number, of reversible Carnot cycles. We can draw a very important conclusion from this fact, namely, that around any complete cycle $\sum (q_{rev}/T)$ or $\oint (dq_{rev}/T)$ is equal to zero. This is analogous to other equations we might write such as $\oint dV = 0$, $\oint dT = 0$, $\oint dE = 0$. These equations imply that the quantities dV, dT, dE, and dq_{rev}/T are exact differentials and that the volume, the temperature, the energy, and something of which dq_{rev}/T is an exact differential are functions of the state of a system. We give the name entropy and the symbol S to the

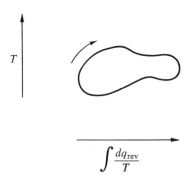

FIGURE 2-5 A general reversible cycle in which heat is exchanged between systems and surroundings at an infinite sequence of infinitesimally different temperatures. It can be seen that this is just the limit to which the representations in Figures 2-3 and 2-4 lead.

property of which dq_{rev}/T is the differential, and we write the defining equation

$$dS \equiv \frac{dq_{rev}}{T} \tag{2-1}$$

We defer consideration of the significance and applications of Equation 2-1 and the entropy concept until the next and subsequent chapters.

> *PROBLEM 2-2*   By recalling that the area under a curve in which y is plotted against x is equal to $\int y\,dx$ and also by making use of the fact that for any cycle $\sum q = -\sum w$, show that the area included within the curve of Figure 2-5 is equal to the work of one complete cycle.

::

THE THERMODYNAMIC TEMPERATURE SCALE OF WILLIAM THOMSON

In 1848 William Thomson (Lord Kelvin) published a paper entitled "On the Absolute Thermometric Scale founded on Carnot's Theory of the Motive Power of Heat and calculated from Regnault's observations."[3] In it, after mentioning the relatively precise methods of temperature comparison that had been developed, he said:

The theory of thermometry is however as yet far from being in so satisfactory a state . . . . In the present state of physical science, therefore, a question of extreme interest arises: *Is there any principle on which an absolute thermometric scale can be founded?* It appears to me that Carnot's theory of the motive power of heat enables us to give an affirmative answer.

The relation between motive power and heat, as established by Carnot, is such that the *quantities of heat*, and *intervals of temperature*, are involved as the sole elements in the expression for the amount of mechanical effect to be obtained through the agency of heat; and since we have, independently, a definite system for the measurement of quantities of heat, we are thus furnished with a measure for intervals according to which absolute differences of temperature may be estimated.

In the present state of science no operation is known by which heat can be absorbed, without either elevating the temperature of matter, or becoming latent and producing some alteration in the physical condition of the body into which it is absorbed; and the conversion of heat (or caloric) into mechanical effect is probably impossible,* certainly undiscovered.

---

* This opinion seems to be nearly universally held by those who have written on the subject. A contrary opinion, however, has been advocated by Mr. Joule of Manchester; some very remarkable discoveries which he has made with reference to the *generation* of heat by friction of fluids in motion, and some known experiments with magneto-electric machines, seeming to indicate an actual conversion of mechanical effect into caloric. No experiment however is adduced in which the converse operation is exhibited; but it must be confessed that as yet much is involved in mystery with reference to these fundamental questions of natural philosophy.[4]

[3] W. Thomson, *Phil. Mag.* (3) **33**, 313 (1848).

[4] We have reproduced Thomson's footnote in full, in spite of the fact that it is not closely relevant to the matters of our immediate discussion, because of its historical interest. It strongly implies that the Second Law was understood and accepted by Professor

. . . The amount of mechanical effect to be obtained by the transmission of a given quantity of heat, through the medium of any kind of engine in which the œconomy is perfect, will depend, as Carnot demonstrates, not on the specific nature of the substance employed as the medium of transmission of heat in the engine, but solely on the interval between the temperatures of the two bodies between which the heat is transferred . . . .

The characteristic property of the scale which I now propose is, that all degrees have the same value; that is, that a unit of heat descending from a body *A* at the temperature *T* of this scale, to a body *B* at the temperature (*T* − 1), would give out the same mechanical effect, whatever be the number *T*. This may justly be termed an absolute scale, since its characteristic is quite independent of the physical properties of any specific substance.

The Kelvin temperature scale embodies this proposal. By Carnot's principle the $-w/q_1$ of 1-59 must have the same value for any Carnot-cycle device operating reversibly between the same pair of temperatures. The relation of the temperature to this efficiency factor is rather arbitrary, but it is convenient and conventional to define the relation to be the same as that found between the efficiency factor and the ideal-gas temperature scale, namely Equation 1-59. Every effort should be made to squelch the easily-come-by notion that in adopting this convention we are somehow still basing our result fundamentally on the properties of gases. The converse is true. The uncomplicated properties of ideal gases in terms of Equations 1-11, 1-12, and 1-38 give us the simplest way (and a very precise way at ordinary temperatures) of evaluating the efficiency factor, and it turns out, of course, that the temperature scale which we provisionally identified as the ideal-gas scale is the Kelvin scale.

$$\frac{-w}{q_1} = \frac{T_1 - T_2}{T_1} \quad \begin{bmatrix} \text{Definition of Kelvin} \\ \text{temperature scale} \end{bmatrix} \qquad (2\text{-}2)$$

PROBLEM 2-3   Equation 2-2 can be rearranged to

$$-w = q_1 \frac{T_1 - T_2}{T_1} \qquad (2\text{-}3)$$

For a very small temperature interval we can write, where *dT* is the increment between the first temperature and the second,

$$\frac{dw}{q} = \frac{dT}{T} \qquad (2\text{-}4)$$

Show that if Thomson had interpreted his own proposal literally—"that a unit of heat descending from a body A at the temperature *T'* of this scale to a

---

Thomson and many of his scientific colleagues before the First Law was. And it also indicates that Thomson deserves much of the credit for rescuing James Joule from disrepute in the scientific community and, by giving careful attention to his work, for hastening general acceptance of the First Law.

body B at the temperature $(T' - 1)$, would give out the same mechanical effect, whatever be the number T'''—that on this basis a temperature T' on such a scale would be related to a temperature T on the Kelvin scale by

$$T' = \ln T$$

On such a scale, then, the coldest temperature would not be $0°$ but $-\infty°$. This result has some significance in relation to the Third Law.

∷

A REVERSIBLE CARNOT CYCLE ON AN ELECTROCHEMICAL SYSTEM

The reversible Carnot cycle which is easiest to analyze is certainly the ideal-gas expansion engine. In the foregoing paragraphs it has been shown that the very general experimental observations that spontaneous processes do not reverse themselves lead to faith in the Second Law. Carnot's principle is a consequence (or a statement) of the Second Law. The fact that some other kind of system undergoing a reversible Carnot cycle must have the same efficiency as an ideal gas engine needs no specific testing. It is instructive, however, to examine an electrochemical example.

Many galvanic cells can be operated under conditions under which the chemical reactions involved occur very nearly reversibly. Such a cell is the one in which the reaction is that between chlorine and silver

$$\text{Ag(c)} + \tfrac{1}{2}\text{Cl}_2\text{(g)} \rightarrow \text{AgCl(c)} \qquad (2\text{-}5)$$

We subject this cell to a constant-pressure, reversible Carnot cycle in which the adiabatic steps are infinitesimal.[5]

[5] It should be understood that when we are considering a reversible chemical or electrochemical reaction, the reaction can go in either direction, and an equilibrium or electrochemical balance might be indicated by using a pair of arrows pointing in opposite directions. However, in order to be able to write equilibrium constant expressions without ambiguity as to what is in the numerator and what is in the denominator, or to write expressions for energy changes or changes in electrical potential without ambiguity as to sign, it seems advantageous to use just one arrow to indicate the direction in which the reaction is being considered to go.

It perhaps should be pointed out that we have chosen as our example a cell in which a mole of reaction is also an electrochemical equivalent of reaction. This simplifies the notation of our derivation. After the derivation has been completed, the result will be generalized to cover cases in which the cell reaction as written corresponds to other than one faraday.

One more fact about this particular kind of cell also facilitates our derivation. Provided that there is enough Ag, Cl_2, and AgCl originally present so that none of these will be completely depleted as a result of the amounts of electrochemical reaction which occur as a result of the passage of the number of faradays assumed in the Carnot cycle considered, and if the pressure of Cl_2 is kept constant, the cell's potential at any particular temperature will be a constant independent of the number of faradays which have passed.

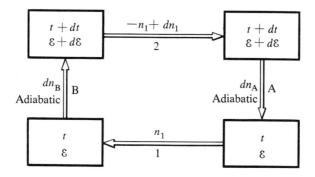

FIGURE 2-6   A Carnot cycle on a galvanic cell. The n's are the number of faradays passed through the cell.

Figure 2-6 represents this cycle. We proceed to analyze this cycle in a manner similar to that employed in Problems 1-27 and 1-28 in which we analyzed the reversible Carnot cycle on an ideal gas. The only new tool we need is the expression for electrical work corresponding to the operation of such a cell. In infinitesimal form it is

$$dw_{el} := -\mathscr{F}\varepsilon\, dn \tag{2-6}$$

where ε is the emf of the cell, dn is the number of faradays passed reversibly through the cell in the direction of the cell reaction as written, and $\mathscr{F}$ is Faraday's constant (23.063 kcal volt^{-1} equiv^{-1}). ε is given a positive sign if the cell reaction tends to occur spontaneously in the direction written, a negative sign otherwise.

Where we begin our cycle is immaterial to the results of our calculations. There is an experimental advantage, however, to starting with an adiabatic step. The cycle is:

A. An infinitesimal adiabatic reversible step in which dn_A faradays are passed through the cell and the temperature falls infinitesimally. (The n's are always counted in the direction of the cell reaction as written above; thus if current flows through the cell in the opposite direction, the value of n or dn is negative.)

1. A finite, isothermal, reversible step in which n_1 faradays pass through the cell and the cell absorbs an amount of heat q.

B. A second infinitesimal, adiabatic, reversible step in which dn_B faradays pass through the cell and the temperature rises infinitesimally.

2. A second finite, isothermal, reversible step in which $-n_1 + dn_1$ faradays pass through the cell—just enough current to restore the cell to its initial state—and the cell gives out an amount of heat $-q'$.

Faraday's law says that the amount of electrochemical reaction is directly proportional to the number of faradays passed through the cell. Then, because at the end of the cycle the cell must be in exactly the same state as it was at the start, the net number of faradays passed must be zero. This argument leads to the equation

$$-dn_1 = dn_A + dn_B \tag{2-7}$$

PROBLEM 2-4   From the First Law, the definition of H and the expression for expansion work show that we can write for steps 1 and 2

$$\Delta H = q + w_{el} \tag{2-8}$$

For an infinitesimal step

$$dH = dq + dw_{el} \tag{2-9}$$

::

The cycle is at constant pressure and hence the net PV work is zero. Therefore, we use w without a subscript to mean electrical work. From Equations 2-3 and 2-6 we can then write

$$w_1 = -n_1 \mathscr{F} \varepsilon \tag{2-10}$$

$$w_2 = (n_1 - dn_1)\mathscr{F}(\varepsilon + d\varepsilon) \tag{2-11}$$

$$dw_A = dH_A \tag{2-12}$$

$$dw_B = dH_B \tag{2-13}$$

It is perhaps worth pointing out that for constant pressure processes in which electrical work is involved, $\Delta H \neq q$ and $\Delta H_{\text{Adiabatic}} \neq 0$.

From time to time we find it convenient to specify the increment in some quantity which corresponds to the amount of reaction indicated by a certain chemical equation. We use the tilde to identify this kind of increment. For example, with reference to 2-5, $\Delta \tilde{H}$ means the ΔH for the reaction between 1 mole of Ag and $\frac{1}{2}$ mole of Cl_2 to give 1 mole of AgCl. Using this notation and bearing in mind that q is the heat absorbed in step 1, we get from 2-8 and 2-10

$$q = n_1 \Delta \tilde{H} + n_1 \mathscr{F} \varepsilon \tag{2-14}$$

Having laid the foregoing groundwork, we now show how cell 2-3 can be used to establish a thermodynamic temperature scale. It is convenient for us now to take the differential Equation 2-4 as the definition of the scale

$$\frac{dw}{q} = \frac{dT}{T} \tag{2-4}$$

PROBLEM 2-5   Noting that dw is the net work around the cycle, and using expressions derived above, show that

$$dw = n_1 \mathcal{F} \, d\varepsilon \tag{2-15}$$

and that

$$\frac{\mathcal{F} \, d\varepsilon}{\Delta \tilde{H} + \mathcal{F}\varepsilon} = \frac{dT}{T} \tag{2-16}$$

∷

In order to avoid the use of too many n's we wrote our cell reaction to correspond to one faraday. Equation 2-16 can be generalized for a reaction in which $\tilde{n}$ electrochemical equivalents are involved by writing

$$\frac{\tilde{n}\mathcal{F} \, d\varepsilon}{\Delta \tilde{H} + \tilde{n}\mathcal{F}\varepsilon} = \frac{dT}{T} \tag{2-17}$$

Although it is not our immediate purpose, it can be pointed out that we have derived a special form of the Gibbs–Helmoltz equation. This form is usually written

$$\tilde{n}\mathcal{F}\left(\frac{d\varepsilon}{dT}\right)_P = \frac{\Delta \tilde{H} + \tilde{n}\mathcal{F}\varepsilon}{T} \tag{2-18}$$

We can indicate a definite integration of 2-17. Thus

$$\ln \frac{T_2}{T_1} = \tilde{n}\mathcal{F} \int_{\varepsilon_1}^{\varepsilon_2} \frac{d\varepsilon}{\Delta \tilde{H} + \tilde{n}\mathcal{F}\varepsilon} \tag{2-19}$$

In Figure 2-7 the necessary data relative to the Ag–Cl$_2$ cell have been plotted to enable us to evaluate 2-19. The area under the curve between any pair of temperatures should be the value of the natural logarithm of the ratio of the two temperatures. In the example T_1 is the triple point of water (273.16 °K by definition), T_2 is the melting point of AgCl (728 °K), and T_3 is 1000 °K.[6]

In the foregoing sections the Second Law has been given quantitative expression for reversible processes in Equations 1-59, 1-60, and 2-4. All the thermodynamic relations that apply to reversible processes and to systems at equilibrium can be built on the foundation already developed. Although we have given the quantity dq_{rev}/T the symbol dS, we have not found it necessary to make any use of this symbol even in such an important derivation as that of the Gibbs–Helmholtz equation.

---

[6] The values used for ΔH are based on the National Bureau of Standards Circular 500 (1952) (for ΔH_{298}) and the work of K. K. Kelley and colleagues reported in Bulletin 584 of the U.S. Bureau of Mines. The emf data are based on the calculated values of Hamer, Malmberg, and Rubin reported in *J. Electrochem. Soc.*, **103**, 8 (1956). Because the ε's were not determined potentiometrically but were calculated from thermal data in a way requiring a prior knowledge of Kelvin temperatures, our example is artificial.

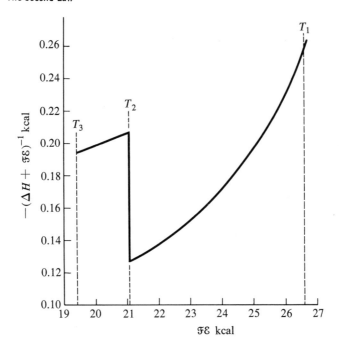

FIGURE 2-7   Plot of $-(H + F\mathscr{E})^{-1}$ vs $F\mathscr{E}$ for reaction 2-5.

The point is that there is much leeway in the manner in which the structure of thermodynamics can be built up. We have already found, for example, that it is more convenient at times to have available a single building block $H(= E + PV)$ than always to have to use the smaller parts E and PV separately. So it turns out with respect to a number of other defined quantities, the free energy and the chemical potential, for example. Of course, not only does the use of these quantities facilitate one's own thermodynamic calculations, but it is necessary to know the language in order to understand and use the work of others.

Yet elaborate and varied as the superstructure may be, all thermodynamics rests on the same very few basic postulates.

ADDITIONAL PROBLEMS

PROBLEM 2-6   Kitchen refrigerators have been designed to use a gas flame (rather than electricity) as the source of the necessary energy to produce the refrigeration. Suppose that the hottest part of the machine is at 300 °C, that heat can be given to the surroundings at 50 °C and that the temperature of the refrigerating coils is −10 °C. Considering Carnot cycles, what is the minimum number of kilocalories of fuel that must be burned to take a kilocalorie of heat from the cold chamber of the refrigerator?

::

PROBLEM 2-7   By estimating the area under the curve of Figure 2-7 calculate the temperature at which the emf of the Ag–Cl$_2$ cell is 1.00 volt. Interpolation of the Hamer, Malmberg, and Rubin tables gives about 270 °C.

::

PROBLEM 2-8   At 25 °C for the reaction

$$AgCl + H_2 \rightarrow HCl(g) + Ag$$

$\Delta H° = -8.30$ kcal. At 25 °C the reversible emf of a cell which corresponds to the above reaction is 0.145 volt. By treating ΔH as constant over the temperature range involved and by integrating 2-17 between the appropriate limits, estimate the emf of this cell at 100 °C.

Answer: 0.091 volt.

3

Spontaneity and
Equilibrium

Probably almost everyone knows what is meant by a *spontaneous process* and by a *state of equilibrium*. It is seldom wasted effort, however, to examine basic expressions such as these; in fact it is perhaps very important to inquire about names which one feels free to use glibly, names about which one might off-hand feel the least need to inquire.

Perhaps the simplest case is that of the isolated system. Here a spontaneous process is any process which occurs. Usually, moreover, one speculates about processes which might occur and would say that if there are adequate grounds for the assertion that such a process is possible—not prohibited by the composition and state of the system and applicable natural laws—then such a process is a spontaneous process. One is usually sufficiently tolerant, too, to permit little triggering manipulations such as the passage of a very small spark to ignite a combustible gas mixture or the snipping of a fine thread to permit two reactants to come together, the releasing of a sensitive detent to let a piston go, or the closing of a delicate switch to complete an electric circuit.[1]

A significant observation about spontaneous processes in isolated systems is that nature is reliable: the reverse of a spontaneous process in an isolated system never occurs. For the isolated system, at least, a non-spontaneous process never occurs.

Equilibrium in the isolated system corresponds to a state from which there can be no (spontaneous) macroscopic changes. Any changes in

---

[1] In this respect our meaning of spontaneous process differs from the usual import of the term spontaneous combustion (alleged sometimes to occur in untidy piles of greasy rags). We would view the burning of the rags as a spontaneous process even if it were started with a match.

such a system would have to be produced by external intervention in excess of the triggering sort, and such intervention would negate the isolation of the system.

Commonly systems are in equilibrium with respect to some kinds of changes but not with respect to others. An example is a mixture of oxygen, hydrogen, and water at room temperature. This system comes quickly to temperature, pressure, and solubility equilibrium but remains far from chemical equilibrium. Another example is a galvanic cell connected to an electric capacitor. The system quickly approaches a state of chemical and electric balance—a kind of constrained equilibrium. The constraint is that the reactants and products are prevented from direct contact and can react only through the electrochemical process. The constraint in this case is never perfect. Diffusion processes do occur and do permit direct chemical interaction to some extent. But thermodynamic treatment of well-behaved cells on an equilibrium basis is quite satisfactory in spite of the imperfection of the constraint and the collateral occurrence of some irreversible or nonequilibrium processes.

Some of the preceding considerations apply to spontaneity and equilibrium in nonisolated systems. But what is meant by a spontaneous change in a system which is not isolated? We encounter some difficulty in giving an acceptable answer to that question. We could avoid the difficulty by avoiding the use of the term in cases of nonisolated systems. But its use is rather wide-spread, for example in the statement that at constant temperature and pressure and where no work other than expansion work is involved, a spontaneous change is accompanied by a decrease in free energy.[2] Therefore, we try to find a satisfactory meaning to give to the word "spontaneous" in connection with nonisolated systems.

Perhaps the best definition of a spontaneous process states that it is any process which can occur under the specified external conditions and allowed system-surroundings interactions. To explore this definition we ask some specific questions. Which are the spontaneous processes: the expansion of a gas or the contraction of a gas, the evaporation of a liquid or the condensation of a vapor, the charging of a battery or the discharging of it?

We would have to give these answers. It is a spontaneous process for a gas to expand if the external pressure is less than that of the system (gas) and with pressure volume interactions permitted, but if the external pressure is initially greater than that of the system, it is a spontaneous process for the system to contract. It is a spontaneous process for a liquid to evaporate if the applied pressure is less than the equilibrium vapor pressure of the liquid but it is also a spontaneous process for the vapor to condense if the applied pressure is greater than the equilibrium pressure. It is a spontaneous process for a battery to become discharged if the potential applied to it by the surroundings is less than a certain value but it is also a spontaneous process

---

[2] The definition of free energy is given later.

for the battery to become charged if the externally applied potential is greater than the reversible potential of the cell.

Yet one may feel that a spontaneous process is one that goes on without help from the surroundings. Few would deny, however, that at atmospheric pressure and at 1 °C the melting of ice is a spontaneous process. Nonetheless, this process gets help from the surroundings not only in the heat which is absorbed but in the PV work (a very small amount, to be sure) that the surroundings do on the system.

We arrive at essentially the same conclusions about the interacting system as we did about the isolated system. Spontaneous processes or changes are any that can occur under the prescribed conditions and allowed interactions. Nonspontaneous processes will not occur. What one might think of as a nonspontaneous process is just one that has been made spontaneous by an alteration in the external conditions or allowed interactions. Equilibrium in the nonisolated system represents a state from which there can be no change without an alteration of the surroundings or the constraints.

But what about the reversible process which must occur under equilibrium conditions? The answer is that the perfectly reversible process is an ideal which can only be approached. The closer it is approached, the smaller becomes the tendency for any change to occur, the driving force of the process approaches zero. So any macroscopic changes which occur in the universe occur by irreversible processes because they will not occur in the absence of a driving force. The term spontaneous process has the connotations of a real process, of a driving force, and of irreversibility.

4

Entropy

CLAUSIUS' INEQUALITY

Equation 2-1 defines the entropy increment.

$$dS = \frac{dq_{rev}}{T} \tag{2-1}$$

We found that for any cycle

$$\oint \frac{dq_{rev}}{T} = 0 \tag{4-1}$$

This implies that S, the entropy, is a function of the state of a system and that ΔS for a prescribed change can have only one value; it cannot have different values for different ways of accomplishing the change, not even if the change is accomplished irreversibly. The restriction of Equation 2-1 to the reversible case means that on the basis of the Second Law we can make quantitative determinations of entropy increments only through measurements on reversible processes.

The qualitative relation between dq_{rev}/T and dq_{irrev}/T is very significant and easily derived. This relation is known as the inequality of Clausius. Consider a specified infinitesimal change in state occurring at T. For this change there is a certain energy increment dE which must have the same value for any process of carrying out the change. We note that an irreversible process is always one in which the work, w, is greater than w for a reversible process.

$$w_{irrev} > w_{rev} \tag{4-2}$$

We can easily satisfy ourselves that this is true whether we are considering a process in which w is positive or whether one in which w is negative. In the first case (surroundings doing work on the system), more work is done on the system to produce the change in the irreversible case because some of it is used in dissipative processes. For the case in which w is negative (the system doing work on the surroundings) less work is done on the surroundings in an irreversible process and therefore the value of w is a smaller negative number. In either case, therefore, 4-2 is valid.[1]

But

$$dq_{irrev} + dw_{irrev} = dE = dq_{rev} + dw_{rev} \qquad (4\text{-}3)$$

Combining 4-2 and 4-3 gives

$$dq_{irrev} < dq_{rev} \qquad (4\text{-}4)$$

which we can divide through by T to get

$$\frac{dq_{irrev}}{T} < \frac{dq_{rev}}{T} = dS \qquad (4\text{-}5)$$

It is the dq/T which is different for the reversible and irreversible cases, not the dS. It is convenient to have a compact expression for the relation between dS and dq/T. This is

$$\frac{dq}{T} \leqq dS \qquad (4\text{-}6)$$

Here, as it has been made clear, the inequality applies to the irreversible process and the equality to the reversible process. With this understanding and also that S is a function of state, 4-6 can be considered a complete statement of the Second Law.

It should be noted that 2-1 (which is the same as the equality in 4-6) is a differential equation and can be used to calculate only values of ΔS not absolute values of S. But this is no more of a limitation than that imposed upon energy calculations by the defining equation 1-7. It is shown that a basis exists for assigning values to the absolute entropies of specified systems. This basis does not exist, however, in the First and Second Laws.

SECOND-LAW ENTROPY CALCULATIONS

As has already been asserted, nothing basically new has been produced by naming dq_{rev}/T the entropy and by giving it the symbol dS. We could plod

---

[1] Customarily one includes on the system side of the system-surroundings boundary any dissipative effects (irreversibility) that are to be taken into account. For example, some authors write for the work of an expansion process which may or may not be reversible,

$$dw = -P_{opp}\, dV$$

where P_{opp} means the pressure of the surroundings (opposing pressure); this for the irreversible case is not equal to the pressure within the system.

our way through thermodynamics always writing dq_{rev}/T, just as we could do without the name energy and the symbol ΔE by always writing $q + w$. But not only would the notation be cumbersome, but the fact is that the quantity represented by dq_{rev}/T or dS and that represented by $q + w$ or ΔE are very basic entities and not just two of the numerous possible combinations of the measurables T, q, and w. We have already seen evidence of this fact in the result that E and S are functions of state. The following examples should further our understanding of the significance of entropy and lead to formulations of its relation to the tendency of systems to change and to move toward states of equilibrium.

For any adiabatic process 4-6 becomes

$$dS \geqq 0 \quad \left[\begin{array}{c} \text{Any adiabatic} \\ \text{process} \end{array} \right] \tag{4-7}$$

Any process that can occur in an isolated system is an adiabatic one. Moreover, because strictly reversible processes have no tendency to occur we can write for any spontaneous process in an isolated system

$$dS > 0 \quad \left[\begin{array}{c} \text{Any spontaneous} \\ \text{process in an} \\ \text{isolated system} \end{array} \right] \tag{4-8}$$

A system without surroundings is an isolated system. In that sense, the universe is an isolated system, and in view of 4-8 we can appreciate Clausius' concise statement of the First and Second Laws: "The energy of the universe is a constant; the entropy of the universe tends always toward a maximum."

Some of the quantitative calculations of entropy changes are simple. One such is the calculation of ΔS for a change in phase such as melting or vaporization at constant temperature. Here

$$\Delta S = \frac{q}{T} = \frac{\Delta H}{T} \quad \left[\begin{array}{c} \text{Reversible phase change at} \\ \text{constant pressure} \end{array} \right] \tag{4-9}$$

In 4-9 T is the temperature at which the phases are in equilibrium under the existing constant pressure and q or ΔH is the heat absorbed in the phase change. It is only a bit more complicated to get an expression for ΔS for changing the temperature of a system of heat capacity C. In this case the heat absorbed at any particular temperature is $C \, dT$ and

$$\Delta S = \int_{T_1}^{T_2} C \frac{dT}{T} \tag{4-10}$$

In 4-10 C is the heat capacity of the system under whatever conditions it is being heated, constant pressure or constant volume, for example. The system need not be simple. It could be one in which chemical reactions are occurring provided that they are occuring reversibly. A restriction on the use of 4-10 is that C must not become infinite between T_1 and T_2 as it would if the melting or boiling point of a substance being heated at constant pressure fell between T_1 and T_2. But that case is easily handled by using 4-9 to calculate the entropies of phase changes and 4-10 to calculate the entropy increments in the temperature intervals between phase changes.

PROBLEM 4-1   Show that when two blocks of equal and constant heat capacity C but at initially different temperatures T_1 and T_2 are put into contact and allowed to come to thermal equilibrium:

$$\Delta S = C \ln \frac{1}{4} \left(\frac{T_1}{T_2} + \frac{T_2}{T_1} + 2 \right) \qquad (4\text{-}11)$$

▪▪

PROBLEM 4-2   We have assumed in solving Problem 4-1 that the blocks exchange heat only between themselves. Together, then, they constitute a system adiabatically separated from surroundings and 4-7 indicates that, the process being irreversible, the value calculated for ΔS must be positive regardless of the values of T_1 and T_2. Prove mathematically that the expression given in answer to Problem 4-1 cannot be a negative number, recognizing that T_1, T_2, and C are always positive numbers.

▪▪

Another typical entropy calculation is that for a change in state of an ideal gas. We consider the most general change possible, namely from $T_1 P_1 V_1$ to $T_2 P_2 V_2$. Figure 4-1 represents this change showing two convenient paths that can be used to make the entropy calculation.

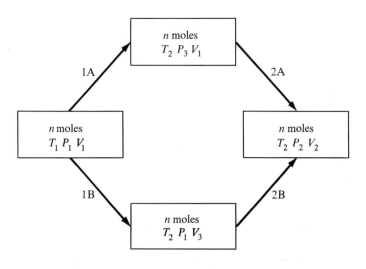

FIGURE 4-1

For step 1A

$$\Delta S = n \int_{T_1}^{T_2} C_V \frac{dT}{T}$$

For step 1B

$$\Delta S = n \int_{T_1}^{T_2} C_P \frac{dT}{T}$$

For the isothermal steps 2A and 2B

$$\Delta E = 0 \quad \text{and} \quad q = -w = \int P \, dV = nR \int \frac{dV}{V} = -nR \int \frac{dP}{P}$$

So for 2A

$$\Delta S = nR \ln \frac{V_2}{V_1}$$

and for 2B

$$\Delta S = nR \ln \frac{P_1}{P_2}$$

We come out with two alternative expressions for ΔS for the specified change, namely,

$$\Delta S = n \int_{T_1}^{T_2} C_V \frac{dT}{T} + nR \ln \frac{V_2}{V_1} \qquad (4\text{-}12)$$

$$\Delta S = n \int_{T_1}^{T_2} C_P \frac{dT}{T} + nR \ln \frac{P_1}{P_2} \qquad (4\text{-}13)$$

PROBLEM 4-3 Certainly the entropy increment cannot be different for different ways of calculating it. For any possible reversible path the result, if correctly calculated, must come out the same. Show that 4-12 and 4-13 are equivalent. This can perhaps be done most easily by subtracting one equation from the other and showing that the difference is zero.

::

PROBLEM 4-4 Calculate the entropy change for an ideal gas expanding into an evacuated space (Joule experiment).

Answer: $\Delta S = nR \ln \dfrac{V_2}{V_1}$

::

Problem 4-4 is a typical example of the calculation of entropy increments accompanying irreversible processes. The procedure is to ignore the irreversible process except to the extent that it is necessary to specify the initial and final states of the system. Once these states are specified one devises any conveniently usable reversible way of making the change in state and evaluates $\sum (dq/T)$ for this reversible path.

FIGURE 4-2

Next let us calculate the entropy of isothermally mixing ideal gases. We assume that not only are the unmixed gases ideal but that they do not interact in the mixture. Specifically, we state our problem in terms of mixing the ideal gases A and B without change in temperature or total pressure or total volume as represented in Figure 4-2.

The direct mixing is an irreversible process. But we can imagine a reversible way of doing the mixing. This would involve the use of semipermeable pistons as indicated in Figure 4-3. The piston with the slanted cross hatching is permeable to A but not to B, the other to B but not to A.

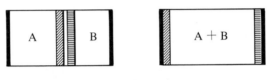

FIGURE 4-3

Thus, the reversible mixing process is equivalent to the reversible isothermal expansion of each gas from its initial volume to the volume of the final mixture. From 4-12 we can write

$$\Delta S = n_A R \ln \frac{V_A + V_B}{V_A} + n_B R \ln \frac{V_A + V_B}{V_B} \tag{4-14}$$

$$\Delta S = n_A R \ln \frac{n_A + n_B}{n_A} + n_B R \ln \frac{n_A + n_B}{n_B} \tag{4-15}$$

$$\Delta S = -R(n_A \ln X_A + n_B \ln X_B) \tag{4-16}$$

where X_A and X_B are the mole fractions in the mixture.

PROBLEM 4-5   Verify by inspection of 4-16 that, as it must in consequence of 4-8, the irreversible mixing of ideal gases is accompanied by a net (system and surroundings) increase in entropy.

As our last example of an entropy calculation in this section we show one way that the entropy change in a chemical reaction can be calculated. Specifically, we consider the reaction

$$Ag + \tfrac{1}{2}Cl_2 \rightarrow AgCl \tag{2-5}$$

carried out isothermally and under a constant pressure. In order to make an entropy calculation we need a reversible way of carrying out this reaction. In Chapter 2 we recognized the fact that an ideal electrochemical process meets our needs. We have, in fact, in Equation 2-9 the expression for q_{rev}. Thus, for the amount of reaction shown in 2-5

$$\Delta S = \frac{\Delta H + \mathscr{F}\varepsilon}{T} \tag{4-17}$$

An obvious rearrangement of 4-17 is

$$\mathscr{F}\varepsilon = T\,\Delta S - \Delta H \tag{4-18}$$

A quantitative definition of the tendency of a reaction to proceed is to some extent arbitrary. This is the same kind of intrinsic arbitrariness that exists in the quantitative definition of a temperature scale. We have seen that even a thermodynamic temperature scale might have various defining equations (see Problem 2-3). However, we know that the tendency of a reaction like 2-5 to proceed increases as the value of the intensive variable ε increases. We also know that in the case that the reversible ε is zero, the cell reaction is in chemical equilibrium. We can then conclude from 4-18 that a system will be in equilibrium with respect to a proposed chemical reaction if for that reaction under conditions of constant temperature and pressure

$$T\,\Delta S - \Delta H = 0 \quad \left[\begin{array}{c} \text{Criterion for} \\ \text{chemical equilibrium} \\ \text{under constant } T \text{ and } P \end{array}\right] \tag{4-19}$$

We can also conclude that under the same conditions any spontaneous chemical reaction requires that

$$T\,\Delta S - \Delta H > 0 \quad \left[\begin{array}{c} \text{Necessary condition} \\ \text{for spontaneous reaction} \\ \text{at constant } T \text{ and } P \end{array}\right] \tag{4-20}$$

ADDITIONAL PROBLEMS

PROBLEM 4-6 Two sticky balls are, just before impact, at 300 °K which is the temperature of the surrounding air. They are moving toward each other and undergo a head-on, completely inelastic collision so that their total kinetic energy, which before impact was 1 kcal, is reduced to zero. The heat capacity of each ball is 1 kcal deg^{-1}. For the system (the two balls) calculate: (*a*) ΔS immediately after impact, and (*b*) ΔS after the balls have returned to 300 °K through loss of heat to the air. (*c*) Calculate the total ΔS (system and surroundings after the balls have returned to 300 °K.

(*d*) If the heat capacity of the balls is 0.5 cal g^{-1} deg^{-1}, calculate the speed of the balls relative to each other before impact.

::

PROBLEM 4-7 A glass ampoule containing 1 mole of liquid water at 100 °C is inside an evacuated container of volume V. The ampoule is broken, the system being maintained at 100 °C. Calculate ΔS for the change in state: (*a*) for the case that $V = 10$ liters, and (*b*) for the case that $V = 100$ liters. Make any reasonable approximations.

::

PROBLEM 4-8 At the cool water tap the pressure is equivalent to a head of 100 feet of water. Make a rough estimate of the entropy produced by drawing a glass of water.

Answer: $\frac{1}{4}$ joule deg^{-1}, (neglecting any effort required to turn the valve on and off, and so on).

::

PROBLEM 4-9 A leaky tank contains originally 10 moles of air at 10 atm and 300 °K. (*a*) Calculate the entropy that has been produced when the pressure in the tank has dropped to 1 atm (isothermally) by leakage into the atmosphere. (*b*) How would the result differ if the gas in the tank had been hydrogen?

Answer: (*a*) 45.8 cal deg^{-1}.

::

PROBLEM 4-10 For the reaction

$$CO + \tfrac{1}{2}O_2 \rightarrow CO_2$$

at 25 °C, $\Delta S° = -20.74$ cal deg^{-1}. From the data in Problem 1-39 calculate the entropy increments (ΔS's) for heating CO, O$_2$, and CO$_2$ from 25 °C to 1000 °C. Then calculate $\Delta S°$ for the reaction at 1000 °C.

5

Criteria of Spontaneity and Equilibrium

IDEAL SURROUNDINGS

As long as the sun continues to rise and set, in fact until long after the last ray of starlight has dimmed and gone out, the entropy of the universe will continue to increase. At least so it is with the universe known to thermodynamics. Therefore the total entropy increment of a system and all its surroundings accompanying any change will always be positive.

On the other hand, many spontaneous changes—the isothermal freezing of a supercooled liquid, for example—are accompanied by a decrease in the entropy of the system. Thus, neither the entropy change of the system alone, except the isolated system, nor the entropy change of the system and all of its real surroundings can be used as an indicator of the possibility of a specified change in a system or of the possible fact that the system is at equilibrium with respect to changes conceivable under existing external conditions and allowed system–surroundings interactions.

Yet the most meaningful macroscopic interpretation of the entropy increment—and a very useful one indeed—is in terms of its fundamental relation to the driving force of change.

To make this fact clear let us reexamine the idea of system and surroundings. The great advantage of dividing the universe into system and surroundings is that we do not have to try to keep track of what goes on beyond the system–surroundings boundary. The system cannot be affected in any way by conditions which exist beyond the boundary. It is true, of course, that conditions at the boundary may be related to those beyond. But it is only the conditions at the boundary that can influence the system, and how these conditions may have been arrived at is entirely irrelevant to the behavior of the system. As far as the behavior of the system is concerned, therefore, we can replace the real surroundings by any kind of imaginary ones which will provide the system with the same boundary conditions; conditions such as pressure, temperature, electrical potential, and any others that are relevant to the allowed system–surroundings interactions.

For our present purposes the ideal surroundings are ones which are of uniform (although not necessarily constant) temperature and which deal reversibly with any forms of energy which they receive from or give to the system. We then isolate the system and these ideal surroundings from everything else. The combination of system and ideal surroundings we call an *isolated complex*. Figure 5-1 is a representation of an isolated complex.

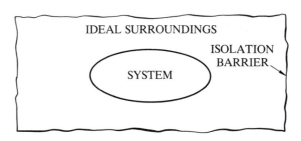

FIGURE 5-1   The isolated complex.

Inequality 4-7 is applicable to the complex. Thus, the system will have the capacity for a specified change if for that change $dS > 0$ for the complex. If $dS = 0$ for the complex, the system is in equilibrium with its surroundings with respect to that change—there is no tendency for the change to occur. A change in which dS for the complex is negative is impossible.

The general expression for dS_{Net} (the entropy increment of the complex) is, of course, the sum of dS_{System} and $dS_{\text{Surroundings}}$. It should be clear that if dq represents the heat absorbed by the system then $-dq$ represents the heat absorbed by the surroundings. Using 2-1 we get, then, (because the ideal surroundings absorb heat reversibly)

$$dS_{\text{Net}} = dS_{\text{System}} - \frac{dq}{T_{\text{Surroundings}}} \tag{5-1}$$

But

$$dS_{\text{Net}} \geqq 0 \tag{4-7}$$

Hence our general criteria for spontaneity and for equilibrium can be written

$$\frac{dq_{\text{System}}}{T_{\text{Surroundings}}} - dS_{\text{System}} \leqq 0 \tag{5-2}$$

where the inequality corresponds to any change which the system can undergo under the specified external conditions (i.e., for which there is a driving force, or which is spontaneous), and the equality corresponds to changes with respect to which the system is in equilibrium with its surroundings. To recall our preceding argument, 5-2 and the above-stated interpretation of its significance must be equally valid when applied to real surroundings which present to the system the same boundary conditions that the ideal surroundings have been assumed to present.

Inasmuch as T is always a positive number, an obvious recasting of 5-2 is

$$dq - T\,dS \leqq 0 \quad \begin{bmatrix} \text{General criteria} \\ \text{for spontaneity} \\ \text{and equilibrium} \end{bmatrix} \quad (5\text{-}3)$$

The cases of most usual interest are ones in which the temperature of the system is maintained the same as that of the surroundings. But, although in 5-3 we have dropped the subscript notation which appears in 5-2, it should be borne in mind that in 5-3 dS refers to the entropy increment of the system and T is the temperature of the surroundings at the system-surroundings boundary whether or not there is at any time a difference between the temperature of the system, or any parts of it, and that of the surroundings at the system–surroundings boundary.[1]

ΔA AND ΔG

A number of sections of the preceding discussions have indicated the importance in thermodynamic calculations of isothermal processes. These sections include, among others, those concerning thermochemistry and those dealing with processes which are not isothermal but which would be very hard to analyze except by breaking them up into isothermal and nonisothermal parts (the Carnot cycle, for example). We now proceed to reduce the general expressions in 5-3 to ones convenient for use in isothermal cases.[2]

---

[1] The practical importance of considering the possibility (implicit in some of the preceding statements) that the system and surroundings have different temperatures at their mutual boundary is with respect to cases in which the system is originally isolated from its surroundings and is then put into contact with them.

[2] One might ask why we turned the inequality sign around in going from a combination of 4-7 and 5-1 to 5-2, and why we take 5-3 rather than 5-2 as the basis of the following development. The answer is in the fact, already stated, that the formal structure of thermodynamics is subject to many arbitrary choices. By using 5-3 instead of 5-2 we obtain a notation which is as sound and as useful as any other; in fact, one which is probably more useful than any other because it is the notation underlying a majority of the literature of thermodynamics and it is a language known to all thermodynamicists. As examples of variants from the formalisms we use, however, we can cite the Planck and Massieu functions (see E. A. Guggenheim, *Thermodynamics, an Advanced Treatment for Chemists and Physicists*, fifth edition, North Holland Publishing Company, Amsterdam, 1967) and the *affinity* function used by de Donder and his school (see Theophile de Donder and P. Van Rysselberghe, *The Thermodynamic Theory of Affinity*, Stanford University Press, 1963 and see I. Prigogine and R. Defay, *Chemical Thermodynamics*, translated by D. H. Everett, Longmans, Green and Co., 1954).

For isothermal cases we can write 5-3 for a finite change

$$q - T \Delta S \leqq 0 \qquad (5\text{-}4)$$

For the two most commonly considered isothermal cases 5-4 becomes

$$\Delta E - T \Delta S \leqq 0 \quad \text{[No-work case]} \qquad (5\text{-}5)[3]$$

$$\Delta H - T \Delta S \leqq 0 \quad \begin{bmatrix} \text{Constant } P \\ \text{no work other} \\ \text{than } PV \text{ work} \end{bmatrix} \qquad (5\text{-}6)[3]$$

PROBLEM 5-1   Starting with 5-4 and other useful relationships, show that 5-5 and 5-6 are valid for the cases indicated. Show also that if system–surroundings interactions are limited to T and PV ones, 5-5 and 5-6 can correctly be asserted to represent the criteria for spontaneity and equilibrium in the constant T and V case and in the constant T and P cases, respectively.

::

It is noteworthy that 4-19 and 4-20 which we derived in a rather specialized way as criteria for equilibrium and spontaneity in chemical reactions are just applications of the generally-derived criteria, 5-6.

All of the quantities appearing in the left-hand members of expressions 5-5 and 5-6 are properties or increments in properties; these members as a whole must therefore represent increments in properties. Inasmuch as 5-5 and 5-6 represent important criteria, it is useful to give names and symbols to these composite properties. We write

$$A \equiv E - TS \quad \begin{bmatrix} \text{Definition of} \\ \text{work content} \end{bmatrix} \qquad (5\text{-}7)[4]$$

and

$$G \equiv H - TS \quad \begin{bmatrix} \text{Definition of} \\ \text{free energy} \end{bmatrix} \qquad (5\text{-}8)[4]$$

---

[3] A large majority of the problems classically dealt with in thermodynamics involve only pressure–volume and temperature interactions between system and surroundings. Authors, including the author of this book, do not always attach a little sign to expressions such as 5-5 and 5-6 indicating whether or not the expression is valid only for cases in which system–surroundings interactions are so limited. It would probably be cumbersome to do so. But the fact that it is not always done requires that the competent thermodynamicist have more than a cookbook knowledge of the subject and, in particular, that he assure himself of the validity of any formulas for cases in which he proposes to use them. This validity should be routinely questioned in cases involving other than T and PV interactions—in cases involving electrical work, for example.

[4] Possibly with a zeal to get into the meat of the matter, readers often pass over the preface to a book. It is suggested that at this point the reader of this book consult the preface, particularly with reference to the comments there on variants of names and symbols of thermodynamic quantities.

PROBLEM 5-2 Show that the following expressions can be obtained by combining 5-3, 5-7, and 5-8

$$dA < 0 \quad \begin{bmatrix} \text{Criterion for spontaneity:} \\ T \text{ constant, no work} \end{bmatrix} \qquad (5\text{-}9)$$

$$dA = 0 \quad \begin{bmatrix} \text{Criterion for equilibrium:} \\ T \text{ constant, no work} \end{bmatrix} \qquad (5\text{-}10)$$

$$dG < 0 \quad \begin{bmatrix} \text{Criterion for spontaneity:} \\ \text{constant } T \text{ and } P, \text{ no work} \\ \text{other than } PV \text{ work} \end{bmatrix} \qquad (5\text{-}11)$$

$$dG = 0 \quad \begin{bmatrix} \text{Criterion for equilibrium:} \\ \text{constant } T \text{ and } P, \text{ no work} \\ \text{other than expansion work} \end{bmatrix} \qquad (5\text{-}12)$$

∷

PROBLEM 5-3 Using 5-7 derive an expression for the maximum work that the system can *do on* the surroundings, $(-w)_{\text{Max}}$, in any isothermal process. This amount of work would correspond to a reversible change. The expression is

$$(-w)_{\text{Max}} = -\Delta A \qquad (5\text{-}13)$$

which is equivalent to

$$\Delta A = w_{\text{Min}} \qquad (5\text{-}14)$$

5-13 and 5-14 are the basis for the names work content and maximum work function which are given to A, the latter name being somewhat more appropriate to the system of thermodynamic notation in which w is counted positive for work done by system on surroundings and in which the first law is written $\Delta E = q - w$.

∷

All of the relationships of the state functions A and G have to be obtained primarily through the defining Equations 5-7 and 5-8 together with First and Second Law relationships of E, H, T, and S to other variables such as pressure, volume, concentration, and the electrical potential. To derive such relationships it is usually necessary to consider infinitesimal increments. From 5-7 and 5-8 we get

$$dA = dE - T\,dS - S\,dT \qquad (5\text{-}15)$$

and

$$dG = dH - T\,dS - S\,dT \qquad (5\text{-}16)$$

or

$$dG = dE + P\,dV + V\,dP - T\,dS - S\,dT \qquad (5\text{-}17)$$

∷

PROBLEM 5-4   For a system that undergoes an infinitesimal change that may involve both PV work and electrical work we have $dw = -P\,dV + dw_{el}$. Show that for a reversible change of this kind at constant T and P

$$dG = dw_{el} \quad \begin{bmatrix} \text{For reversible change at} \\ \text{constant } T \text{ and } P \text{ with} \\ \text{no work other than } PV \\ \text{and electrical work} \end{bmatrix} \quad (5\text{-}18)$$

In the light of 2-6 this can be written

$$dG = -\mathscr{F}\varepsilon\,dn \tag{5-19}$$

for the reaction in a reversibly-operating voltaic cell. Here dn is the number of faradays corresponding to dG.

⠿

The solution to Problem 5-4 illustrates the use of two commonly useful manipulations in derivations of that kind. One is to set $dE = dq + dw$. The other is to set $dq - T\,dS = 0$ for a reversible process.

For equilibria of the kind previously mentioned in which only the pressure and temperature of the surroundings influence the system we can, by setting $dq = T\,dS$ and $dw = -P\,dV$, derive some useful relations from the defining equations for A and G (through 5-15 and 5-17).

PROBLEM 5-5   Derive the following equations:

$$dA = -P\,dV - S\,dT \tag{5-20}$$

$$dG = V\,dP - S\,dT \tag{5-21}$$

$$\left(\frac{\partial A}{\partial V}\right)_T = -P \tag{5-22}$$

$$\left(\frac{\partial A}{\partial T}\right)_V = -S \tag{5-23}$$

$$\left(\frac{\partial G}{\partial P}\right)_T = V \tag{5-24}$$

$$\left(\frac{\partial G}{\partial T}\right)_P = -S \tag{5-25}$$

⠿

PROBLEM 5-6   Show that when an ideal gas expands isothermally

$$\Delta G = nRT \ln \frac{P_2}{P_1} \tag{5-26}$$

Is this equation valid whether or not the gas is expanded reversibly?

⠿

THE GIBBS–HELMHOLTZ EQUATION

Although A and G are state functions useful chiefly in calculations about isothermal processes, it is important to know how the value of ΔG (or ΔA) for a chemical reaction carried out isothermally at one temperature is related to the value of ΔG for the same reaction carried out isothermally at a different temperature.[5]

By applying to the cycle shown in Figure 5-2 the principle that ΔG for any system undergoing any cycle must be zero, we can readily derive

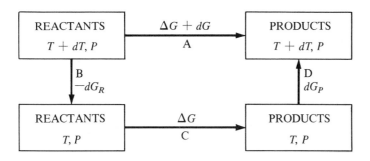

FIGURE 5-2   The change in G for a chemical reaction resulting from a change in the temperature at which the reaction is carried out.

the Gibbs–Helmholtz equation. The equation expressing this principle for the series of steps shown in Figure 5-2 is

$$\Delta G + d\,\Delta G = -dG_R + \Delta G + dG_P \tag{5-27}$$

where ΔG is the ΔG for the reaction carried out at T and dG_R and dG_P are the increments in G for reactants and products respectively when these are heated from T to $T + dT$.

PROBLEM 5-7   Verify Equation 5-27.

⁖

To evaluate dG_R and dG_P we start with 5-22. Noting that the pressure is constant we get

$$dG_R = -S_R\,dT \quad \text{and} \quad dG_P = -S_P\,dT \tag{5-28}$$

---

[5]Inasmuch as G is a state function, the value of ΔG for specified initial and final states of the system cannot depend on the process whereby the change in state occurs. In the present context, then, an isothermal reaction means only that the initial temperature of the reactants is the same as the final temperature of the products.

and 5-27 becomes

$$d \, \Delta G = -(S_P - S_R) \, dT = -\Delta S \, dT \tag{5-29}$$

or

$$\left(\frac{\partial \, \Delta G}{\partial T}\right)_P = -\Delta S \qquad \left[\begin{array}{c} \text{Gibbs–Helmholtz} \\ \text{equation} \end{array}\right] \tag{5-30}^6$$

PROBLEM 5-8   Verify the statement in Chapter 2 that 2-18 is a special form of the Gibbs–Helmholtz equation.

∷

In solving Problem 5-8 we have made use of Equations 2-6, 5-18, and 5-19 together with 5-30. In view of the fact that 5-21 which led to 5-30 was based on the assumption of only P and T interactions, why is it legitimate to use 5-30 as the basis for getting an expression (2-18) which evidently is related to electrical systems? The answer is that the $\partial \varepsilon / \partial T$ in 2-18 is a partial derivative with both T and n held constant. Here n is the number of faradays that have passed through the cell (it is to be distinguished from the reaction parameter $\tilde{n}$ which is the number of faradays corresponding to the amount of reaction indicated by the reaction equation as written). Thus if n is held constant during a heating process, no current is being passed through the cell and the cell is not interacting electrically with its surroundings during this process. We can take these things into account by recasting 2-18 as

$$\tilde{n} \mathcal{F} \left(\frac{\partial \varepsilon}{\partial T}\right)_{n,P} = \frac{\Delta H + \tilde{n} \mathcal{F} \varepsilon}{T} \tag{5-31}$$

PROBLEM 5-9   Use the defining equation for G together with 5-30 to show that

$$\left(\frac{\partial \, \Delta G}{\partial T}\right)_P = \frac{\Delta G - \Delta H}{T} \tag{5-32}$$

∷

PROBLEM 5-10   Verify the following equations

$$\frac{\partial \left(\dfrac{\Delta G}{T}\right)}{\partial T} = \frac{1}{T} \frac{\partial \, \Delta G}{\partial T} - \frac{\Delta G}{T^2} \tag{5-33}$$

and

$$\frac{\partial \left(\dfrac{\Delta G}{T}\right)}{\partial \left(\dfrac{1}{T}\right)} = \Delta G - T \frac{\partial \, \Delta G}{\partial T} \tag{5-34}$$

∷

---

[6] The P in this equation indicates a partial derivative at constant pressure; it should not be confused with the P used in the foregoing derivation to refer to products.

PROBLEM 5-11   Hence show that

$$\frac{\partial\left(\frac{\Delta G}{T}\right)}{\partial T} = \frac{-\Delta H}{T^2} \tag{5-35}$$

and

$$\frac{\partial\left(\frac{\Delta G}{T}\right)}{\partial\left(\frac{1}{T}\right)} = \Delta H \tag{5-36}$$

∷

The great practical importance of the various forms of the Gibbs–Helmholtz equations such as 5-30, 5-31, 5-35, and 5-36 lies in the possibilities they offer for evaluating certain important parameters (properties) from measurements of others. This is, of course, the great practical importance of thermodynamics itself. Our first use of a Gibbs–Helmholtz relationship in this way was in showing that 2-17 is an equation we can use to evaluate absolute temperatures by calorimetric and potentiometric measurements on galvanic cells. Other kinds of applications should be apparent. If ΔG can be determined as a function of T then 5-30, 5-35, and 5-36, for example, enable us to calculate values for ΔH and ΔS. Conversely, if ΔH or ΔS is known as a function of T, then 5-30 or 5-35 and 5-36 can be integrated to permit the evaluation of ΔG at any temperature if its value at any one temperature is known.

PROBLEM 5-12   Suppose that for a certain chemical reaction ΔH is correctly given by an equation of the form of 1-57. Using 5-35 show that the value of ΔG at any temperature T, ΔG_T, is correctly represented by

$$\frac{\Delta G_T}{T} = \frac{\Delta G_{T'}}{T'} + \frac{1}{T}(\Delta H_{T''} + \Delta A) - \Delta a \ln T - \tfrac{1}{2}\Delta bT$$

$$\tfrac{1}{6}\Delta cT^2 - \frac{1}{2}\frac{\Delta d}{T^2} + B \tag{5-37}$$

In this equation T' is a temperature for which ΔG is known and T'' is a temperature for which ΔH is known. ΔA and B are constants for a particular reaction (not the ΔA of 5-7).

∷

PROBLEM 5-13   Write expressions for ΔA and B in terms of T', T'' and other symbols appearing in 2-37. For many reactions standard enthalpies and free energies of formation of reactants and products are known and tabulated for $25\,°C$ ($298.15\,°K$). For such cases one would solve numerical problems by setting $T' = T'' = 298.15°$.

∷

STABILITY AND EQUILIBRIUM

Kirkwood and Oppenheim[7] give a rather detailed mathematical treatment of the stability of various kinds of equilibria. Their discussion of this topic is based on Gibbs' treatment[8] contained in his most outstanding contribution to thermodynamics: *On the Equilibrium of Heterogeneous Substances.* Although we do not explore this topic as extensively as the cited authors, there are reasons that we should not ignore it. These reasons have to do with the preciseness of statements and the carefulness of interpretation which are required in such an exact and elegant science as thermodynamics.

As a starting point let us take some statements which one may encounter in textbooks. One may read that for a conceivable change from state a to state b in an isolated system:

If ΔS is positive, the reaction tends to proceed from state a to state b.

If ΔS is zero, the system is at equilibrium and no spontaneous process occurs.

If ΔS is negative, the reaction tends to go spontaneously in the reverse direction, i.e., from b to a.

Again, of the free energy in a constant P and T change one may read that:

If ΔG is negative, the process tends to proceed spontaneously.

If ΔG is zero the system is at equilibrium.

If ΔG is positive, the process tends to proceed in the opposite direction.

Let us question the unqualified validity of the statements about the negative ΔS case and about the positive ΔG case.

First, consider two blocks A and B in thermal contact with each other and isolated from everything else. Suppose that A and B are originally at the same temperature. A transfer of heat from A to B will correspond to a net decrease in entropy. But so will a transfer of heat from B to A, and the process cannot occur in either direction.

Second, consider the electrochemical system of Figure 5-3. Suppose that this system, isolated electrically from its surroundings but maintained at constant T and P, has come to electrochemical balance. If now

[7] J. G. Kirkwood and Irwin Oppenheim: *Chemical Thermodynamics*, McGraw-Hill Book Company, Inc., New York, 1961.

[8] J. Willard Gibbs: *Collected Works*, Longmans, Green and Co., Inc., New York, 1928. In paperback the first edition (1906) of the Longmans, Green collection is published (1961) by Dover Publications, Inc., New York.

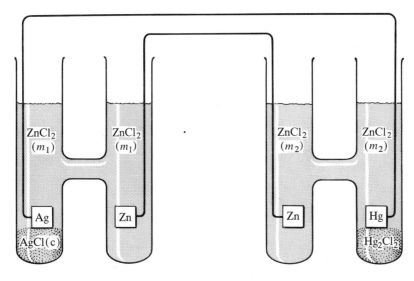

FIGURE 5-3

we consider the possibility of a flow of current through the system in a certain direction we find that this current flow would correspond to an increase in the free energy of the system. But we could show that a flow of current in the opposite direction would also correspond to an increase in free energy; thus neither a forward nor a reverse change of this kind can occur under the stated conditions.

The preceding two are examples of stable equilibria. There is a barrier to any change from a uniquely describable equilibrium state. Neutral equilibrium, on the other hand, is illustrated by the equilibrium between a liquid and its vapor at a specified temperature. In this case the volume of the system may be changed within limits (the limits being the volumes at which the vapor phase and the liquid phase, respectively, disappear) without disturbing the equilibrium. Similar examples of neutral equilibrium can be readily thought of which involve heterogeneous chemical reactions.

In view of the preceding discussion of stable equilibria, we may well question our assertions that for the isolated system the criterion of equilibrium with respect to a conceivable infinitesimal change is that dS for the change is zero, and that for a system maintained at constant T and P the criterion of equilibrium with respect to a conceivable infinitesimal change is that dG for the change is zero. The facts can be clarified by reference to a parameter which measures the extent to which the change has occurred, the *degree of advancement* of the change. In the case of a transfer of heat, this parameter might well be taken to be q itself. In the case of an electrochemical process the degree of advancement parameter could logically be taken to be the number of coulombs or faradays which have flowed through the circuit;

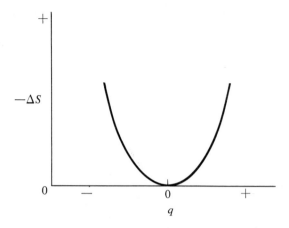

FIGURE 5-4 S as a function of q for the transfer of heat between two blocks initially at the same temperature.

and for a chemical reaction a good choice for this parameter would be the number of moles of a particular reactant that have reacted.

Figures 5-4, 5-5, and 5-6 are plots of ΔS or ΔG against an appropriate degree of advancement parameter for the three examples of equilibrium that we have discussed. If we use the symbol ξ for the advancement parameter in general, it can be seen that the respective criteria of equilibrium are

$$\frac{\partial S}{\partial \xi} = 0 \quad \begin{bmatrix} \text{Isolated} \\ \text{system} \end{bmatrix} \tag{5-38}$$

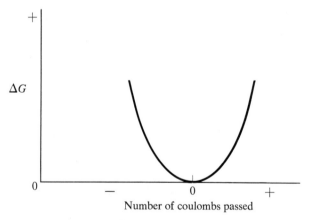

FIGURE 5-5 G as a function of the number of coulombs passed through the system of Figure 5-3.

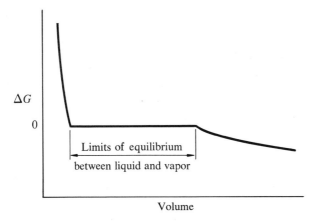

FIGURE 5-6 *G* as a function of the volume of a liquid-vapor system maintained at constant temperature.

and

$$\frac{\partial G}{\partial \xi} = 0 \qquad \begin{bmatrix} \text{Constant } T, P \text{ and} \\ \text{only } T \text{ and } PV \\ \text{interactions} \\ \text{with surroundings} \end{bmatrix} \qquad (5\text{-}39)$$

The distinction between the neutral equilibrium case and the stable equilibrium cases is that for stable equilibrium

$$\frac{\partial^2 S}{\partial \xi^2} < 0 \qquad \begin{bmatrix} \text{Stable equilibrium in an} \\ \text{isolated system} \end{bmatrix} \qquad (5\text{-}40)$$

$$\frac{\partial^2 G}{\partial \xi^2} > 0 \qquad \begin{bmatrix} \text{Stable equilibrium in a constant } T \\ \text{and } P \text{ system which has only } T \\ \text{and } PV \text{ interactions with} \\ \text{its surroundings} \end{bmatrix} \qquad (5\text{-}41)$$

whereas for the neutral equilibrium cases

$$\frac{\partial^2 S}{\partial \xi^2} = 0 \qquad \begin{bmatrix} \text{Neutral equilibrium in an} \\ \text{isolated system} \end{bmatrix} \qquad (5\text{-}42)$$

$$\frac{\partial^2 G}{\partial \xi^2} = 0 \qquad \begin{bmatrix} \text{Neutral equilibrium in a system at} \\ \text{constant } T \text{ and } P \text{ with only } T \\ \text{and } PV \text{ interactions with} \\ \text{surroundings} \end{bmatrix} \qquad (5\text{-}43)$$

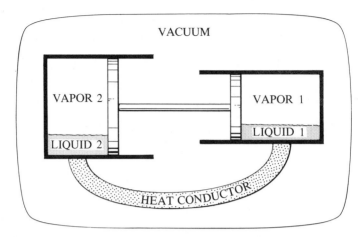

FIGURE 5-7 An isolated system which, with suitable choice of substances, cylinder bores, and operating temperature range, can exhibit neutral equilibrium. A random choice of these parameters would be expected to lead to a state of stable equilibrium.

Figure 5-7 shows an isolated system comprising two opposing piston-cylinder combinations. Different liquids, each in equilibrium with its vapor, are contained in the two cylinders and the bores of the cylinders are so adjusted that at the temperature of the apparatus the forces on the opposing pistons are exactly equal. This isolated system will exhibit neutral equilibrium if at the existing temperature the molar heats of vaporization of the two liquids are equal. If they are unequal, there will be stable equilibrium.[9]

Figure 5-8 shows two soap bubbles connected by a tube. If the radii and surface tensions of the two bubbles are equal, the system will be in

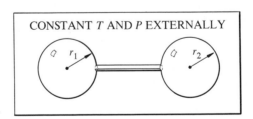

FIGURE 5-8 Two soap bubbles connected to permit gas to flow between them. This system will be in unstable equilibrium when $r_1 = r_2$.

[9] These assertions are based on certain approximations such as that the vapors are ideal gases. These approximations are pointed out later in connection with derivation of the Clausius–Clapeyron equation.

unstable equilibrium. The free energy as a function of the difference in radii is plotted qualitatively in Figure 5-9. The free energy equations applicable to this type of equilibrium are

$$\frac{\partial G}{\partial \xi} = 0 \quad \begin{bmatrix} \text{Unstable equilibrium at constant } T \text{ and } P \\ \text{with only } T \text{ and } PV \text{ interactions} \\ \text{between system and surroundings} \end{bmatrix} \quad (5\text{-}44)$$

$$\frac{\partial^2 G}{\partial \xi^2} < 0 \qquad\qquad\qquad\qquad\qquad\qquad\qquad\qquad (5\text{-}45)$$

It is sometimes more difficult but also more rewarding to find the truth in a statement than to adduce grounds for denying its truth. Of course the truth is often established by attempts to deny it. This is the basis

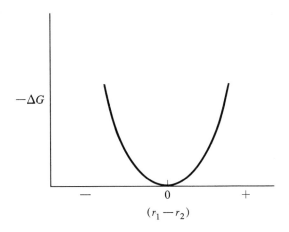

FIGURE 5-9

we have set forth for the First and Second Laws. Thus, instructive as it has hopefully been, our criticisms of the statements about the negative ΔS in the isolated system and about the positive ΔG in the system at constant T and P rest on thin ice. When we pointed out that the flow of heat in either direction between blocks initially at the same temperature would correspond to a decrease in entropy, we were considering two *different* changes in state, not the same change in state proceeding in opposite directions. In one case the final temperature of A would be higher and of B lower than the original temperature. The reverse of this change in state *would* correspond to a positive ΔS and *would* be spontaneous. In the other case the final temperature of B would be higher and of A lower than the original temperature and the reverse of this change in state would be spontaneous.

ADDITIONAL PROBLEMS

PROBLEM 5-14   From data in Table 1-1, Problem 1-39 and Problem 4-10 calculate for the reaction

$$CO + \tfrac{1}{2}O_2 \rightarrow CO_2$$

(*a*) $\Delta G°$ at 25 °C, (*b*) $\Delta G°$ at 1000 °C.

⁝

PROBLEM 5-15   It follows immediately from 5-24 that at constant temperature for an equilibrium process

$$\Delta G = \int V \, dP$$

Consider a liter of water, supercooled to -1.00 °C which completely fills a strong, nonexpandable container. The water is initially at 1 atm and is kept isothermal at -1.00 °C throughout the experiment. A spontaneous process sets in and ice begins to form. However very little can form because ice and water are almost noncompressible and the sturdy container prevents an overall volume increase. The pressure rises to about 120 atm which is the pressure at which ice and water are in equilibrium at -1.00 °C. Note that the work done in this process is zero.

By considering a two-step reversible process whereby the change in state could be brought about (increasing the pressure on the supercooled liquid to 120 atm, then freezing the small amount of it that freezes in the spontaneous process), calculate ΔG for this spontaneous process.

Answer: $\Delta G = +2.9$ kcal. Are you surprised that ΔG is positive?

6

Thermodynamic Formulas

The eight state functions P, V, T, E, H, S, A, G have a certain preeminence in thermodynamics. A secondary echelon is composed of the first partial derivatives of these functions with respect to each other. For example, the two common kinds of heat capacity we have shown to be represented by the derivatives

$$C_V = \left(\frac{\partial E}{\partial T}\right)_V \tag{1-29}$$

and

$$C_P = \left(\frac{\partial H}{\partial T}\right)_P \tag{1-30}$$

The thermal expansion coefficient α and the isothermal compressibility β are defined

$$\alpha = \frac{1}{V}\left(\frac{\partial V}{\partial T}\right)_P \tag{6-1}$$

$$\beta = -\frac{1}{V}\left(\frac{\partial V}{\partial P}\right)_T \tag{6-2}$$

It is to be noted that these partial derivatives have definite values only under equilibrium conditions. That is, if one were to increase the pressure on an elastic fluid extremely rapidly, the temperature and pressure would momentarily be nonuniform throughout the fluid and the volume would not, in the first instants, have reached its final value. Inasmuch, then, as these partial derivatives are understood to correspond to infinitesimal changes carried out under equilibrium conditions (reversibly) we can set $dq = T\,dS$ and $dw = -P\,dV$, and we can indicate an adiabatic process by indicating that S is constant. Thus, for adiabatic compressions the partial derivatives such as

$(\partial T/\partial P)_S$ and $(\partial P/\partial V)_S$ would be of interest. All of these partial derivatives are also state functions; they are properties of systems.

Some of the second partial derivatives would commonly be of practical interest also. One might be interested, for example, in the variation in the heat capacity of a system with temperature or its variation with pressure. These variations would be denoted by second partial derivatives such as

$$\left(\frac{\partial C_V}{\partial T}\right)_V = \left(\frac{\partial^2 E}{\partial T^2}\right)_V \tag{6-3}$$

or

$$\left(\frac{\partial C_P}{\partial T}\right)_P = \left(\frac{\partial^2 H}{\partial T^2}\right)_P \tag{6-4}$$

and

$$\left(\frac{\partial C_V}{\partial P}\right)_T = \frac{\partial}{\partial P_T}\left(\frac{\partial E}{\partial T}\right)_V = \frac{\partial^2 E}{\partial P\,\partial T} \tag{6-5}$$

$$\left(\frac{\partial C_P}{\partial P}\right)_T = \frac{\partial^2 H}{\partial T\,\partial P} \tag{6-6}$$

The first and second partial derivatives are written in the preceding examples in terms of only two independent variables. This implies that the state of the system in question can be uniquely defined by specifying values for any two of the eight thermodynamic functions. Otherwise the constancy of more than one parameter would have to be indicated in the partial derivative symbol.

All the properties of a system will be unique functions of any two of these state functions if the system is homogeneous and if the permitted interactions with the surroundings are limited to thermal and PV interactions. Hence, the methods and results presented in this section are directly applicable only to systems so limited, but by treating each phase separately and by introducing other variables these methods and results can be extended to heterogeneous systems which are subject to other kinds of interactions with the surroundings including interactions involving exchange of substance.

Slater[1] points out in his 16-page summary of classical thermodynamics that there are 336 first partial derivatives and that it can be shown that it is possible to find a relation among any four of these and certain of the thermodynamic variables. This consideration leads to the conclusion that there are 521,631,180 different relations or thermodynamic formulas of this kind.

This number is probably an exaggeration. Slater seems to be counting separately each derivative of pairs such as

$$\left(\frac{\partial P}{\partial T}\right)_V \quad \text{and} \quad \left(\frac{\partial T}{\partial P}\right)_V$$

[1] J. C. Slater, *Introduction to Chemical Physics*, McGraw-Hill Book Co., Inc. New York, 1939.

These are, of course, just reciprocals of one another, and the substitution of

$$\frac{1}{\left(\dfrac{\partial T}{\partial P}\right)_V} \quad \text{for} \quad \left(\frac{\partial P}{\partial T}\right)_V$$

would produce a new formula differing only trivially from the original. But if we take this fact into account and say that there are 168 rather than 336 first partial derivatives we still get a very large number, namely 32,018,910 for the number of thermodynamic formulas involving first partial derivatives but not second or higher derivatives.

Obviously no one is going to have derived or be familiar with more than a minute fraction of all of these thermodynamic equations. In the rest of this chapter we derive a few of them, chiefly as examples of methods that can be used to get them and many others.[2]

A very simple procedure leads to a number of relations. This procedure is to substitute into the basic equations for dE, dH, dA, and dG the expressions $dq = T\, dS$ and $dw = -P\, dV$. First we get

$$dE = T\, dS - P\, dV \tag{6-7}$$

$$dH = dE + d(PV) = T\, dS + V\, dP \tag{6-8}$$

$$dA = dE - d(TS) = -P\, dV - S\, dT \tag{6-9}$$

$$dG = dH - d(TS) = V\, dP - S\, dT \tag{6-10}$$

We can then hold one variable constant in any of the above equations and thereby get an expression for a partial derivative. By holding E constant in 6-7, for example, we get $T\, dS = P\, dV$ at constant E or

$$\left(\frac{\partial S}{\partial V}\right)_E = \frac{P}{T} \tag{6-11}$$

If, instead, in the same equation we hold S constant we get

$$\left(\frac{\partial E}{\partial V}\right)_S = -P \tag{6-12}$$

If we hold V constant we get

$$\left(\frac{\partial E}{\partial S}\right)_V = T \tag{6-13}$$

---

[2] The methods described in the following paragraphs follow closely Slater's treatment. This, in turn, is based on the work of P. W. Bridgman reported in *A Condensed Collection of Thermodynamic Formulas*, Harvard University Press, 1925.

Many authors have presented adaptations and developments of Bridgman's proposals. We cite here two additional references G. N. Lewis and M. Randall, *Thermodynamics*, revised by K. S. Pitzer and L. Brewer, *loc. cit.* (see this reference, Appendix 6, page 665); B. Carroll, "On the Use of Jacobians in Thermodynamics," *J. Chem. Ed.*, **42**, 218 (1965). Carroll points out that the use of Jacobians is particularly advantageous in dealing with systems of which the state is a function of more than two independent variables.

To illustrate another method let us first observe that by a treatment of 6-9 analogous to the preceding we get

$$\left(\frac{\partial A}{\partial V}\right)_T = -P \tag{6-14}$$

and

$$\left(\frac{\partial A}{\partial T}\right)_V = -S \tag{6-15}$$

What is often referred to as the reciprocity relation expresses the fact that in getting a mixed second derivative of a state function, the order of differentiation is immaterial. For example

$$\frac{\partial^2 A}{\partial V \, dT} = \frac{\partial}{\partial T}\left(\frac{\partial A}{\partial V}\right)_T = \frac{\partial}{\partial V}\left(\frac{\partial A}{\partial T}\right)_V \tag{6-16}$$

By combining 6-16 with 6-14 and 6-15 we get the relation

$$\left(\frac{\partial P}{\partial T}\right)_V = \left(\frac{\partial S}{\partial V}\right)_T \tag{6-17}$$

PROBLEM 6-1   By treatments of Equations 6-8, 6-9, and 6-10 in a manner analogous to that used on 6-7 to get 6-11, 6-12, and 6-13, derive nine more expressions for first partial derivatives.

::

PROBLEM 6-2   By use of the reciprocity relation on 6-7, 6-8, and 6-10 get the other three Maxwell's relations analogous to 6-17.

::

Another useful manipulation applicable to exact differentials, which the increments of all state functions are and which dq and dw are not, may be called the cyclic rule. Let us write

$$dx = \left(\frac{\partial x}{\partial y}\right)_z dy + \left(\frac{\partial x}{\partial z}\right)_y dz \tag{6-18}$$

Now set $dx = 0$. We get after rearrangement

$$\left(\frac{\partial x}{\partial y}\right)_z \left(\frac{\partial y}{\partial z}\right)_x \left(\frac{\partial z}{\partial x}\right)_y = -1 \tag{6-19}$$

PROBLEM 6-3   Show that 6-19 follows from 6-18.

::

PROBLEM 6-4   The Joule–Thomson coefficient, μ_{JT}, is

$$\mu_{JT} = \left(\frac{\partial T}{\partial P}\right)_H \tag{1-49}$$

Noting that

$$C_P = \left(\frac{\partial H}{\partial T}\right)_P \tag{1-30}$$

show by the cyclic rule that

$$\left(\frac{\partial H}{\partial P}\right)_T = -C_P \mu_{JT} \tag{6-20}$$

∷

Another simple operation for developing thermodynamic formulas depends upon the fact that the different partial derivatives having the same variable held constant may be manipulated as though they were ordinary fractions. This is a very useful fact because by holding one of the eight thermodynamic functions constant, P for example, and writing the partial derivatives of six of the remaining seven with respect to the seventh, T for example, then by division one can get expressions for all the other fifteen of this kind of partial derivative in terms of the original six. So, for example,

$$\left(\frac{\partial H}{\partial V}\right)_P = \frac{\left(\frac{\partial H}{\partial T}\right)_P}{\left(\frac{\partial V}{\partial T}\right)_P} \tag{6-21}$$

PROBLEM 6-5   Show that for an ideal gas

$$\left(\frac{\partial H}{\partial V}\right)_P = \frac{PC_P}{R} \tag{6-22}$$

∷

A relation between C_P and C_V can be obtained by a derivation slightly more complex than the methods described above. First,

$$C_V = \left(\frac{\partial q}{\partial T}\right)_V = T\left(\frac{\partial S}{\partial T}\right)_V \tag{6-23}$$

and

$$C_P = \left(\frac{\partial q}{\partial T}\right)_P = T\left(\frac{\partial S}{\partial T}\right)_P \tag{6-24}$$

Then, on the basis that dS can be considered a variable dependent on any two of the other state functions we write

$$dS = \left(\frac{\partial S}{\partial T}\right)_V dT + \left(\frac{\partial S}{\partial V}\right)_T dV \tag{6-25}$$

and

$$dS = \left(\frac{\partial S}{\partial T}\right)_P dT + \left(\frac{\partial S}{\partial P}\right)_T dP \tag{6-26}$$

∷

PROBLEM 6-6   By multiplying both 6-25 and 6-26 by T show that we can get two expressions for $T\,dS$ which are

$$T\,dS = C_V\,dT + T\left(\frac{\partial P}{\partial T}\right)_V dV \qquad (6\text{-}27)$$

$$T\,dS = C_P\,dT - T\left(\frac{\partial V}{\partial T}\right)_P dP \qquad (6\text{-}28)$$

(To solve this problem one has to use two of Maxwell's relations. See Problem 6-2.)

::

PROBLEM 6-7   By subtracting 6-27 from 6-28 and setting $dV = 0$ show that the following expressions for $C_P - C_V$ result:

$$(C_P - C_V)\,dT = T\left(\frac{\partial V}{\partial T}\right)_P dP \qquad (6\text{-}29)$$

$$C_P - C_V = T\left(\frac{\partial V}{\partial T}\right)_P\left(\frac{\partial P}{\partial T}\right)_V \qquad (6\text{-}30)$$

$$C_P - C_V = -T\,\frac{\left(\dfrac{\partial V}{\partial T}\right)_P^{\,2}}{\left(\dfrac{\partial V}{\partial P}\right)_T} \qquad (6\text{-}31)$$

::

Inasmuch as it is possible to get a relation among any four of the first derivatives of the thermodynamic variables (which relation may include also these variables themselves) it is possible to express all 168 first derivatives in terms of only three of them. Table 6-1 gives a number of these first derivatives all expressed in terms of the thermal expansion, the compressibility, and the heat capacity properties of systems, namely in terms of $(\partial V/\partial T)_P$, $(\partial V/\partial P)_T$, and C_P. In some cases alternative formulas involving C_V instead of C_P are given. These partial derivatives have been chosen because they are usually relatively easy to measure. Many more expressions are readily obtainable from this table than are actually listed.[3] The expressions that are tabulated were obtained by the methods that have been described in this chapter.

---

[3] Expressions for all 168 first partial derivatives would be directly obtainable by the type of combination illustrated by 6-21 if the table had included a similar set of six partial derivatives with E held constant, a set of six with H held constant, a set of six with A held constant, and a set of six with G held constant.

Table 6-1   Table of Thermodynamic Relations*

$$\left(\frac{\partial V}{\partial T}\right)_P = \left(\frac{\partial V}{\partial T}\right)_P$$

$$\left(\frac{\partial S}{\partial T}\right)_P = \frac{C_P}{T}$$

$$\left(\frac{\partial E}{\partial T}\right)_P = C_P - P\left(\frac{\partial V}{\partial T}\right)_P$$

$$\left(\frac{\partial H}{\partial T}\right)_P = C_P$$

$$\left(\frac{\partial A}{\partial T}\right)_P = -P\left(\frac{\partial V}{\partial T}\right)_P - S$$

$$\left(\frac{\partial G}{\partial T}\right)_P = -S$$

$$\left(\frac{\partial V}{\partial P}\right)_T = \left(\frac{\partial V}{\partial P}\right)_T$$

$$\left(\frac{\partial S}{\partial P}\right)_T = -\left(\frac{\partial V}{\partial T}\right)_P$$

$$\left(\frac{\partial E}{\partial P}\right)_T = -T\left(\frac{\partial V}{\partial T}\right)_P - P\left(\frac{\partial V}{\partial P}\right)_T$$

$$\left(\frac{\partial H}{\partial P}\right)_T = V - T\left(\frac{\partial V}{\partial T}\right)_P$$

$$\left(\frac{\partial A}{\partial P}\right)_T = -P\left(\frac{\partial V}{\partial P}\right)_T$$

$$\left(\frac{\partial G}{\partial P}\right)_T = V$$

$$\left(\frac{\partial P}{\partial T}\right)_V = -\frac{\left(\frac{\partial V}{\partial T}\right)_P}{\left(\frac{\partial V}{\partial P}\right)_T}$$

$$\left(\frac{\partial S}{\partial T}\right)_V = \frac{C_V}{T} = \frac{C_P}{T} + \frac{\left(\frac{\partial V}{\partial T}\right)_P^2}{\left(\frac{\partial V}{\partial P}\right)_T}$$

$$\left(\frac{\partial E}{\partial T}\right)_V = C_V = C_P + T\frac{\left(\frac{\partial V}{\partial T}\right)_P^2}{\left(\frac{\partial V}{\partial P}\right)_T}$$

$$\text{Table 6-1} \quad \text{(Cont.)}$$

$$\left(\frac{\partial H}{\partial T}\right)_V = C_V - V \frac{\left(\frac{\partial V}{\partial T}\right)_P}{\left(\frac{\partial V}{\partial P}\right)_T} = C_P + T \frac{\left(\frac{\partial V}{\partial T}\right)_P^2}{\left(\frac{\partial V}{\partial P}\right)_T} - V \frac{\left(\frac{\partial V}{\partial T}\right)_P}{\left(\frac{\partial V}{\partial P}\right)_T}$$

$$\left(\frac{\partial A}{\partial T}\right)_V = -S$$

$$\left(\frac{\partial G}{\partial T}\right)_V = -S - V \frac{\left(\frac{\partial V}{\partial T}\right)_P}{\left(\frac{\partial V}{\partial P}\right)_T}$$

$$\left(\frac{\partial P}{\partial T}\right)_S = \frac{C_P}{T\left(\frac{\partial V}{\partial T}\right)_P}$$

$$\left(\frac{\partial V}{\partial T}\right)_S = \frac{\left(\frac{\partial V}{\partial P}\right)_T}{\left(\frac{\partial V}{\partial T}\right)_P} \frac{C_V}{T} = \frac{\left(\frac{\partial V}{\partial P}\right)_T}{\left(\frac{\partial V}{\partial T}\right)_P} \left[\frac{C_P}{T} + \frac{\left(\frac{\partial V}{\partial T}\right)_P^2}{\left(\frac{\partial V}{\partial P}\right)_T}\right]$$

$$\left(\frac{\partial E}{\partial T}\right)_S = -P \frac{\left(\frac{\partial V}{\partial P}\right)_T}{\left(\frac{\partial V}{\partial T}\right)_P} \frac{C_V}{T} = -P \frac{\left(\frac{\partial V}{\partial P}\right)_T}{\left(\frac{\partial V}{\partial T}\right)_P} \left[\frac{C_P}{T} + \frac{\left(\frac{\partial V}{\partial T}\right)_P^2}{\left(\frac{\partial V}{\partial P}\right)_T}\right]$$

$$\left(\frac{\partial H}{\partial T}\right)_S = \frac{V C_P}{T\left(\frac{\partial V}{\partial T}\right)_P}$$

$$\left(\frac{\partial A}{\partial T}\right)_S = -P \frac{\left(\frac{\partial V}{\partial P}\right)_T}{\left(\frac{\partial V}{\partial T}\right)_P} \frac{C_V}{T} - S$$

$$= -P \frac{\left(\frac{\partial V}{\partial P}\right)_T}{\left(\frac{\partial V}{\partial T}\right)_P} \left[\frac{C_P}{T} + \frac{\left(\frac{\partial V}{\partial T}\right)_P^2}{\left(\frac{\partial V}{\partial P}\right)_T}\right] - S$$

$$\left(\frac{\partial G}{\partial T}\right)_S = \frac{V C_P}{T\left(\frac{\partial V}{\partial T}\right)_P} - S$$

PROBLEM 6-8 Show that one entry in a table of six partial derivatives with respect to T, G held constant, can be obtained directly from 6-10.

Answer:

$$\left(\frac{\partial P}{\partial T}\right)_G = \frac{S}{V} \tag{6-32}$$

**

PROBLEM 6-9 Obtain a second entry for the table proposed in the preceding problem by the following procedure. First rewrite 6-10 in the form

$$dT = \frac{1}{S}(V\,dP - dG) \tag{6-33}$$

This leads, as in the preceding problem, to

$$\left(\frac{\partial T}{\partial P}\right)_G = \frac{V}{S} \tag{6-34}$$

It also leads to

$$\left(\frac{\partial T}{\partial G}\right)_P = -\frac{1}{S} \tag{6-35}$$

Now, if one gets two expressions for $\partial^2 T/\partial G\,\partial P$ by partial differentiation of 6-34 and 6-35 and sets these two expressions equal to each other (reciprocity relation), Equation 6-36 results

$$\frac{1}{S}\left(\frac{\partial V}{\partial G}\right)_P - \frac{V}{S_2}\left(\frac{\partial S}{\partial G}\right)_P = \frac{1}{S^2}\left(\frac{\partial S}{\partial P}\right)_G \tag{6-36}$$

which can be written

$$\frac{1}{S}\left(\frac{\partial V}{\partial G}\right)_P - \frac{V}{S^2}\left(\frac{\partial S}{\partial G}\right)_P = \frac{1}{S^2}\left(\frac{\partial S}{\partial T}\right)_G\left(\frac{\partial T}{\partial P}\right)_G \tag{6-37}$$

Now solve for $(\partial S/\partial T)_G$ by evaluating all of the other partial derivatives in 6-37 from 6-32 and Table 6-1.

Answer:

$$\left(\frac{\partial S}{\partial T}\right)_G = \frac{C_P}{T} - \frac{S}{V}\left(\frac{\partial V}{\partial T}\right)_P \tag{6-38}$$

**

PROBLEM 6-10 Lumsden points out[4] that it is difficult to measure the isothermal compressibility of molten salts but that the adiabatic compressibility β_S has in some cases been evaluated through its relation to the velocity of sound and by measurement of the velocity of sound in molten salts. The definition of β_S is

$$\beta_S = -\frac{1}{V}\left(\frac{\partial V}{\partial P}\right)_S \tag{6-39}$$

[4] J. Lumsden, *Thermodynamics of Molten Salt Mixtures*, Academic Press, London and New York, 1966.

The isothermal compressibility β is defined by 6-2. Show that

$$\frac{\beta}{\beta_S} = \frac{C_P}{C_V} \tag{6-40}$$

::

PROBLEM 6-11   Show that

$$\frac{C_P}{C_V} = 1 + \frac{\alpha^2 V T}{\beta_S C_P} \tag{6-41}$$

where α is the thermal expansion coefficient (6-1).

::

PROBLEM 6-12   It is sometimes asserted that C_P is larger than C_V because of the expansion work done against the atmosphere in heating a substance at constant pressure. This statement is valid in the case of an ideal gas (why?). At 1 atm and between $0°$ and $4\,°C$ water has a negative expansion coefficient. This means that in heating water between $0°$ and $4\,°C$ at ordinary pressures PV work will be done by the atmosphere on the water. By reference to thermodynamic formulas prove that, nevertheless, in this interval $C_P > C_V$ for water and that in general $C_P > C_V$ for any system to which these formulas apply regardless of the values of any of the properties of the system.

::

PROBLEM 6-13   By combining 6-20 with one of the formulas in Table 6-1 show that the Joule–Thomson coefficient is given by

$$\mu_{JT} = \frac{1}{C_P}\left[T\left(\frac{\partial V}{\partial T}\right)_P - V \right] \tag{6-42}$$

::

PROBLEM 6-14   Derive expressions for μ_{JT} for (a) an ideal gas and (b) a van der Waals' gas, both for the case that $P \to 0$.

Answers: (a) 0, (b) $\dfrac{1}{C_P}\left(\dfrac{2a}{RT} - b\right)$.

::

The answers to this problem seem to involve a paradox. As the volume of a van der Waals' gas is increased, its equation of state may be made to approach as closely as one likes to $PV = nRT$ yet Answer b indicates that the Joule–Thomson coefficient of a van der Waals' gas approaches a constant but in general nonzero value as V is increased without limit. However, Answer a shows that for a gas whose equation of state is $PV = nRT$, $\mu_{JT} = 0$. How can this paradox be explained? The experimental fact is that Joule–Thomson coefficients of gases have approximately constant values at a given temperature and moderate to zero pressure rather than values that approach zero as $P \to 0$.

PROBLEM 6-15 Derive an expression for one more of the partial derivatives of the type $(\partial X/\partial T)_G$. (See Problems 6-8 and 6-9.)

::

We now consider briefly systems which can interact with the surroundings in ways in addition to the usual T and PV interactions. This is not an entirely new topic; we have already considered at some length a system involving electrical interactions and have in 5-31 set down the expression for a partial derivative with two variables held constant.

As a preface to the consideration of a few thermodynamic formulas applicable to these somewhat more complicated systems we write the forms of the energy increment corresponding to several kinds of potentials.

A general treatment would require vector notation: the magnitudes of electric, magnetic, gravitational, and even pressure fields (except the pressure in fluids) depend upon the direction of measurement. However, most practical thermodynamic problems can be treated satisfactorily by considering only one appropriate direction and the magnitude of the field in that direction.

First we write expressions for some work energy increments.

$$dw_{\text{Expansion}} = -P\,dV \tag{1-16}$$

$$dw_{\text{Electric}} = -\mathscr{F}\varepsilon\,dn \tag{6-43}$$

$$dw_{\text{Magnetic}} = \mathscr{H}\,dM \tag{6-44}$$

($\mathscr{H}$ is the magnetic field strength, M is the magnetization.)

$$dw_{\text{Gravitational}} = mg\,dh \tag{6-45}$$

(g is the acceleration of gravity, h is the height.)

$$dw_{\text{Stretching}} = \tau\,dL \tag{6-46}$$

(This is the 1-dimensional analog of 1-16. τ is the tension, L is the length.)

$$dw_{\text{Surface}} = \sigma\,dA_s \tag{6-47}$$

(This is the 2-dimensional analog of 1-16. σ is the surface tension, A_s is the area.)

$$dw_{\text{Chemical}} = (\mu_2 - \mu_1)\,dn \tag{6-48}$$

(μ is the chemical potential, to be defined later. dn is the number of moles of a substance which undergoes a change from an environment in which its chemical potential is μ_1, to one in which its chemical potential is μ_2.)

Thermal energy increments can be written in similar forms.

$$dq = C_V\,dT \quad \text{(Constant-volume change in } T) \tag{6-49}$$

$$dq = C_P\,dT \quad \text{(Constant-pressure change in } T) \tag{6-50}$$

$$dq = \Delta H_{T_r}\,dn \quad \begin{array}{l}\text{(Heat of transition of } dn \text{ moles} \\ \text{in a process such as vaporization)}\end{array} \tag{6-51}$$

$$dq = \Delta\tilde{H}\,d\xi \tag{6-52}$$

(Heat of a chemical reaction at constant P and T. ξ is the degree of advancement of the reaction, $\Delta \tilde{H}$ is the enthalpy increment per unit of degree of advancement)

$$dq = T\,dS \quad \text{(for any reversible process)} \tag{2-1}$$

It can be seen that each of the foregoing expressions for dw and dq are products of an extensive factor and an intensive one, although the differential is in some cases the extensive factor and in others the intensive one.

As was pointed out earlier, formulas of the type being developed and considered in this chapter refer to infinitesimal processes being carried out under equilibrium conditions. For our present purposes, therefore, we may always use 2-1 for the thermal energy increment. The same reason lets us use any of the preceding expressions for dw without inquiring about possible disparities between the values of the intensive variables in the system and their values in the surroundings.

PROBLEM 6-16   For a closed system in which surface effects are to be considered one can write

$$dE = T\,dS - P\,dV + \sigma\,dA_s \quad \begin{bmatrix} PV \text{ and} \\ \text{surface} \\ \text{work only} \end{bmatrix} \tag{6-53}$$

Write the corresponding expressions for dH, dA, and dG. Show that the reciprocity relation applied to the expression for dG gives

$$\left(\frac{\partial S}{\partial A_s}\right)_{T,P} = -\left(\frac{\partial \sigma}{\partial T}\right)_{P,A_s} \tag{6-54}$$

::

PROBLEM 6-17   Using 6-54 and the expression written in Problem 6-16 for dH, show that

$$\left(\frac{\partial H}{\partial A_s}\right)_{T,P} = \sigma - T\left(\frac{\partial \sigma}{\partial T}\right)_{P,A_s} \tag{6-55}$$

::

Although it is possible to deal with any system by adding new variables such as ε, σ, A_s, etc. to the basic set P, V, E, H, S, A, and G, it may be more efficient in some applications to replace or eliminate some of the basic set. Wall[5] points out the fact that there is almost no volume change when rubbers are stretched; the increase in length is almost exactly compensated by the decrease in cross section. To the approximation that this is true and also that PV work on thermal expansion of the rubber is negligibly small,

---

[5] Frederick T. Wall, *Chemical Thermodynamics*, Second Edition, W. H. Freeman and Company, San Francisco and London, 1965. Chapter 15 gives a very good introduction to the classical and statistical thermodynamics of rubber.

one can set up a special thermodynamic notation from which P and V are eliminated and replaced by τ and L and in which the rubber analogs of the conventional properties would be as follows.

$$dE = T\,dS + \tau\,dL \tag{6-56}$$

$$dH' = T\,dS - L\,d\tau \tag{6-57}$$

$$dA = -S\,dT + \tau\,dL \tag{6-58}$$

$$dG' = -S\,dT - L\,d\tau \tag{6-59}$$

E, A, T, and S have the same significance as in the conventional system.

PROBLEM 6-18   Show that

$$\left(\frac{\partial S}{\partial L}\right)_T = -\left(\frac{\partial \tau}{\partial T}\right)_L \tag{6-60}$$

::

PROBLEM 6-19   The heat capacity of rubber at constant length should be accessible to measurement. It would be defined by

$$C_L \equiv \left(\frac{\partial q}{\partial T}\right)_L = T\left(\frac{\partial S}{\partial T}\right)_L \tag{6-61}$$

Hence

$$\left(\frac{\partial S}{\partial T}\right)_L = \frac{C_L}{T} \tag{6-62}$$

By the use of 6-60 and of the cyclic rule show that

$$\left(\frac{\partial \tau}{\partial T}\right)_L = \frac{C_L}{T}\left(\frac{\partial T}{\partial L}\right)_S \tag{6-63}$$

It can be verified by the simple experiment of quickly stretching or un-stretching a rubber band and then quickly touching it to one's face that the temperature rises upon adiabatic stretching and vice versa. This fact together with 6-63 requires that the tension at constant length will increase as T is increased. This fact is also easily verified.

::

Problems 6-18 and 6-19 have illustrated that for systems involving other than PV work it may be advantageous to modify the definitions of the basic thermodynamic functions. We use as another example of this procedure the treatment of a system subject to magnetic as well as temperature and pressure interactions with the surroundings. We have given in 6-44 the expression for magnetic work.[6] Epstein[7] writes as the definition of a generalized Gibbs function (generalized here only with respect to the effects of a magnetic field)

$$\mathscr{G} = E + PV - \mathscr{H}M - TS \qquad (6\text{-}64)$$

PROBLEM 6-20   Using 6-44 and 6-64 show that

$$d\mathscr{G} = V\,dP - S\,dT - M\,d\mathscr{H} \qquad (6\text{-}65)$$

From 6-65 derive 6-66, 6-67, and 6-68

$$\left(\frac{\partial V}{\partial \mathscr{H}}\right)_{P,T} = -\left(\frac{\partial M}{\partial P}\right)_{T,\mathscr{H}} \qquad (6\text{-}66)$$

$$\left(\frac{\partial S}{\partial P}\right)_{T,\mathscr{H}} = -\left(\frac{\partial V}{\partial T}\right)_{P,\mathscr{H}} \qquad (6\text{-}67)$$

$$\left(\frac{\partial S}{\partial \mathscr{H}}\right)_{T,P} = \left(\frac{\partial M}{\partial T}\right)_{P,\mathscr{H}} \qquad (6\text{-}68)$$

∷

PROBLEM 6-21   The closest approaches to $0\,°K$ have been achieved by adiabatic demagnetization. The principle of this method is indicated as follows. First, the heat capacity of a magnetic substance may depend upon the field strength, and we define

$$C_{PM} \equiv \left(\frac{\partial q}{\partial T}\right)_{P,M} = T\left(\frac{\partial S}{\partial T}\right)_{P,M} \qquad (6\text{-}69)$$

$$C_{P\mathscr{H}} \equiv \left(\frac{\partial q}{\partial T}\right)_{P,\mathscr{H}} = T\left(\frac{\partial S}{\partial T}\right)_{P,\mathscr{H}} \qquad (6\text{-}70)$$

Then we write dS as a function of the three independent variables T, P, and $\mathscr{H}$, and note that for an adiabatic (reversible) process $dS = 0$

$$dS = \left(\frac{\partial S}{\partial T}\right)_{P,\mathscr{H}} dT + \left(\frac{\partial S}{\partial P}\right)_{\mathscr{H},T} dP + \left(\frac{\partial S}{\partial \mathscr{H}}\right)_{T,P} d\mathscr{H} = 0 \qquad (6\text{-}71)$$

Show that 6-72 follows from 6-71 and other previously derived relationships

$$C_{P,\mathscr{H}}\frac{dT}{T} - \left(\frac{\partial V}{\partial T}\right)_{P,\mathscr{H}} dP + \left(\frac{\partial M}{\partial T}\right)_{P,\mathscr{H}} d\mathscr{H} = 0 \qquad (6\text{-}72)$$

[6] Pitzer and Brewer (*op. cit.*) give a more detailed treatment using vector fields and taking into account the possibility that the permeability of the medium in which the field exists is other than that of a vacuum.

[7] Paul S. Epstein, *Textbook of Thermodynamics*, John Wiley and Sons, Inc., New York, 1937. Also, as additional reference containing interesting and clear treatments of systems involving special variables, see Vanderslice, Schamp, and Mason, *Thermodynamics* (Chapter 10), Prentice-Hall, Englewood Cliffs, N.J., 1966; see also all of Terrell Hill, *Thermodynamics for Chemists and Biologists*, Addison-Wesley, 1968.

If the pressure remains constant, the relationship between T and $\mathcal{H}$ in reversible adiabatic variations of field strength is

$$\left(\frac{\partial T}{\partial \mathcal{H}}\right)_{P,S} = -\frac{T}{C_{P,\mathcal{H}}}\left(\frac{\partial M}{\partial T}\right)_{P,\mathcal{H}} \tag{6-73}$$

PROBLEM 6-22   Get an expression for $(\partial T/\partial \mathcal{H})_{P,S}$ for a paramagnetic substance which follows Curie's law which is

$$M = \frac{k\mathcal{H}}{T} \quad \begin{bmatrix} \text{Curie's} \\ \text{law} \end{bmatrix} \tag{6-74}$$

in which k is a positive constant. Hence confirm that adiabatic demagnetization of such a material will cool it.

ADDITIONAL PROBLEMS

PROBLEM 6-23   For water at 25 °C the thermal coefficient of expansion α and the coefficient of isothermal compressibility β are $\alpha = 2.60 \times 10^{-4}\,\text{deg}^{-1}$ and $\beta = 4.85 \times 10^{-5}\,\text{atm}^{-1}$. Taking C_P for water as 1.00 cal g^{-1} and the density as 1.00 g ml^{-1} calculate (a) C_P/C_V for water at 25 °C, (b) the coefficient of adiabatic compressibility β_S for water at 25 °C. *Answers:* (a) 1.010   (b) 4.80 × 10^{-5} atm^{-1}

PROBLEM 6-24   The surface area of a soap bubble A_s (inside and outside surfaces) is $8\pi r^2$ and the volume V is $\frac{4}{3}\pi r^3$. Write an expression for dA_s/dV. The work done in increasing the volume of the bubble is equivalent to the work done in increasing the surface area. By using Equations 1-16 and 6-47 for these two kinds of work show that the excess pressure P_x inside the bubble is $P_x = 4\sigma/r$. Does this result explain why the situation represented in Figure 5-8 represents unstable equilibrium? Note that for a spherical droplet of liquid $P_x = 2\sigma/r$.

PROBLEM 6-25   The velocity of sound is given by

$$\text{velocity} = \sqrt{\frac{1}{\beta_S d}}$$

where d is the density of the medium. From the data in Problem 6-23 calculate the velocity of sound in water at 25 °C (be careful about units). *Answer:* 1.46 × 10^5 cm sec^{-1}.
The reverse of this calculation is the basis for determining β_S from measurements of the speed of sound.

PROBLEM 6-26   Calculate the velocity of sound in air at 25 °C treating air as an ideal gas of molecular weight 28.8, and C_V of $\frac{5}{2}R$ mole^{-1}. *Answer:* 3.47×10^5 cm sec^{-1}

::

PROBLEM 6-27   The surface tension of water at 25 °C is 72 dynes cm^{-1}. Calculate the excess pressure in a water droplet of radius 10^{-3} cm.

::

PROBLEM 6-28   A scientist who had a good ear for music had the problem of analyzing mixtures of hydrogen and air. He decided to use an organ pipe as his analytical instrument. In one experiment the result was that when the pipe was blown with the hydrogen–air mixture the pitch was exactly one-half octave higher than when blown with air alone at the same temperature. Making the approximations that the gases are ideal, that the molecular weight of air is 28.8, and that C_P is 7.0 cal deg^{-1} mole^{-1} for air, hydrogen, and the mixture, calculate the mole fraction H_2 in the mixture. *Answer:* 0.54

7

The Practical Third Law

The Third Law of thermodynamics is a very useful principle, but probably no entirely satisfactory statement of it can be made. Its statements will either be subject to exception or else will be limited to a group such as "perfect crystalline substances," the members and nonmembers of which group cannot, in some cases, be identified *a priori* with certainty. Although the problem of borderline cases will not be completely solved from an experimental point of view, our understanding of the Third Law as well as of the Second Law will be much increased when, subsequently, we examine some of the basic relations between microscopic structure and energy distribution and the thermodynamic (macroscopic) properties of matter. For example, the experimentally found residual entropy of water, carbon monoxide, and at least some crystalline solid solutions (see Eastman and Milner, *loc. cit. infra*, for case of AgCl–AgBr) can be made to seem reasonable on the basis of statistical considerations.

For present purposes we give a statement which we call the practical Third Law, using practical to imply that it is a working principle rather than a complete and rigorous summary of the facts with which it deals. We then show the kind of experimental basis which exists for the postulate and how it can be used to obtain much useful thermodynamic information.

Our practical statement of the Third Law is: *At any pressure the entropy of crystalline elements and compounds may be taken to approach zero as the temperature approaches* 0 °K. (7-1)

The language of this statement is intended to exclude glassy solids and solid solutions. There are exceptions to the statement, but relatively few. It is pointed out here that isotopic solid solutions may usually be treated as conforming to the Third Law. We find in Chapter 8, that if we should say that the entropy of crystalline $^{35}Cl_2$ and that of crystalline $^{37}Cl_2$ are each zero at 0 °K, we would have to conclude that the entropy would be greater than zero for a crystalline solid solution containing the statistical distribution of $^{35}Cl_2$, $^{37}Cl_2$, and $^{35}Cl–^{37}Cl$ found in naturally occurring chlorine. But there is no appreciable isotopic separation or fractionation in ordinary chemical reactions. This means that the same "entropy of mixing" would be present in elemental natural chlorine and in all compounds of natural chlorine. Hence, as a practical matter, natural chlorine and its compounds can be considered subject to the above statement of the Third Law; that is, the entropy of mixing of isotopes can be neglected because it cancels out in chemical reactions.

Some authors attempt to eliminate or reduce the number of exceptions to the practical statement of the Third Law by restricting it to "pure" crystalline substances. However, the criticism of the use of pure is like that mentioned above to the use of perfect. For example, chemically pure ice must be assigned a residual entropy at 0 °K whereas chemically pure naturally-occurring chlorine, which is a mixture of isotopes, can, as we have seen, correctly be assigned zero entropy at 0 °K for purposes of any thermodynamic calculations about systems which do not involve isotopic separation.

Several corollaries of the practical Third Law applicable to substances which are subject to it can be readily derived. One is that the values of ΔH and ΔG for chemical reactions among crystalline substances approach equality (inasmuch as $\Delta S \to 0$) as the temperature approaches zero. In fact experimental evidence pointing to this result doubtless contributed to the formulation of the Third Law principle.

Another corollary follows from the independence of the entropy at 0 °K of the pressure. We can write immediately

$$\left(\frac{\partial S}{\partial P}\right)_T \to 0 \quad \text{as} \quad T \to 0 \tag{7-2}$$

But volume may be considered a function of P and T and hence it also follows that

$$\left(\frac{\partial S}{\partial V}\right)_T \to 0 \quad \text{as} \quad T \to 0 \tag{7-3}$$

PROBLEM 7-1   By the methods of Chapter 6 derive

$$\left(\frac{\partial S}{\partial P}\right)_T = -\left(\frac{\partial V}{\partial T}\right)_P \tag{7-4}$$

and hence that the thermal expansion coefficient of crystals should approach zero as $T \to 0$.

$$\left(\frac{\partial V}{\partial T}\right)_P \to 0 \quad \text{as} \quad T \to 0 \tag{7-5}$$

This result has been verified experimentally.

PROBLEM 7-2   Show that

$$\left(\frac{\partial P}{\partial T}\right)_V \to 0 \quad \text{as} \quad T \to 0 \tag{7-6}$$

::

PROBLEM 7-3   The postulate that the entropy of crystalline elements and compounds remains finite at temperatures above $0\,°\text{K}$ is sometimes included in the statement of the Third Law. Show that this postulate leads to the conclusion (which is verified by experiments) that

$$C_V \to 0 \quad \text{as} \quad T \to 0$$

and   (7-7)

$$C_P \to 0 \quad \text{as} \quad T \to 0$$

::

The experimental procedure by which the Third Law can be verified is indicated in broad outline in Figure 7-1. ΔS_R and ΔS_P are

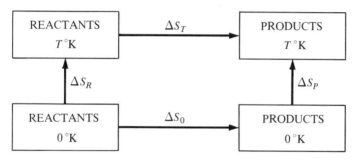

FIGURE 7-1

determined by calorimetric measurements (almost always at constant pressure), the results of which are then treated by

$$\Delta S = \int \frac{C_P\, dT}{T} \quad \begin{bmatrix} \text{For temperature} \\ \text{intervals in which} \\ \text{no phase changes} \\ \text{occur} \end{bmatrix} \tag{4-10}$$

and

$$\Delta S = \frac{\Delta H}{T} \quad \begin{bmatrix} \text{For phase} \\ \text{transitions} \end{bmatrix} \tag{4-9}$$

and must be extrapolated to $0\,°\text{K}$.

ΔS_T is determined by a Second-Law method. The usual procedure would be to write from 5-8 for the constant temperature case

$$\Delta S = \frac{1}{T}(\Delta H - \Delta G) \tag{7-8}$$

Then ΔH is measured calorimetrically and ΔG is evaluated through equilibrium experiments (measurement of the dissociation pressure of an oxide, for example) or measurements on electrochemical cells. Alternatively, ΔS_T can be evaluated more directly in some cases through the Gibbs–Helmholtz equation, 5-30, applied to an electrochemical cell operating reversibly at constant T and P which becomes

$$\Delta S = n\mathscr{F}\frac{d\varepsilon}{dT} \tag{7-9}$$

or through the Gibbs–Helmholtz equation applied to the measured variation in the equilibrium constant with temperature.

It is clear from Figure 7-1 that

$$\Delta S_0 = \Delta S_R + \Delta S_T - \Delta S_P$$

For cases in which reactants and products are crystalline at $0\,°K$, our practical statement of the Third Law is supported by showing that $\Delta S_0 = 0$. It should be brought out that there is something arbitrary about assigning zero as the value of the entropy of crystals at $0\,°K$. All that experiments of the type outlined above can do is to show that the difference in entropy between reactants and products at $0\,°K$ is almost always zero within the limits of accuracy of experiment and calculation. This fact is substantially the postulate proposed by Nernst.[1]

We take as our example the same reaction we have used before, namely

$$\text{Ag} + \tfrac{1}{2}\text{Cl}_2 \rightarrow \text{AgCl} \tag{2-5}$$

First we calculate ΔS_P and ΔS_R.

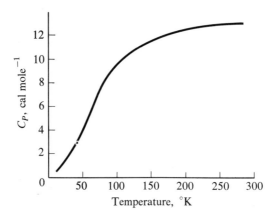

FIGURE 7-2   Heat capacity of AgCl.

[1] Walther Nernst, Nachr. kgl. Ges. Wiss. *Göttingen Math.-physik. Kl.*, **1906**, 1.

Probably the simplest calculation is that of ΔS_P, i.e., of the entropy increment of AgCl from $0°$ to 298.15 °K. Figure 7-2 represents the data of Clusius and Harteck[2] and those of Eastman and Milner.[3] These data are in good agreement with those of several other investigators (cited by Clusius and Harteck) who did not carry the measurements to such low temperatures.

In order to calculate ΔS_P we need to evaluate the integral in 4-10 from $0°$ to 298 °K. The most obvious way to do this graphically would be to plot C_P/T against T. But 4-10 is equivalent to

$$\Delta S = \int C_P \, d \ln T \tag{7-10}$$

which means that the graphical integration can also be done on the basis of a plot of C_P vs $\ln T$. Table 7-1 gives the values of some representative points from the data of Clusius and Harteck and those of Eastman and Milner.

Table 7-1   Heat Capacity of Silver Chloride

(cal deg^{-1} mole^{-1})

$T °K$	C_P	$T °K$	C_P
10.50	0.446*	88.56	9.501
12.64	0.792*	114.0	10.296*
15.00	1.122	114.16	10.493
18.13	1.698*	125.6	10.600*
21.07	2.259	149.13	11.203
25.90	3.110*	174.28	11.582
28.70	3.623	200.24	11.876
43.04	5.735	250.36	12.182
63.3	7.982*	292.12	12.080

* Data of Clusius and Harteck. Data without asterisks are those of Eastman and Milner.

PROBLEM 7-4   Plot the two graphs: C_P/T vs T and C_P vs $\ln T$ for AgCl. It is seen that the latter is less curved and therefore possibly more convenient for estimating areas.

::

By using a large graph or by carrying out a numerical integration with the help of a computer, the integrals can be evaluated over the experimental range with all the precision available in the data. The data must always be extrapolated, however, from the lowest experimental temperatures to 0 °K. The Debye–Sommerfeld equation is

$$C_V = aT + bT^3 \tag{7-11}$$

[2] K. Clusius and P. Harteck, *Zeit. fur physik Chemie*, **A134**, 243 (1928).
[3] E. D. Eastman and R. T. Milner, *J. Chem. Phys.*, **1**, 444 (1933).

This equation was derived theoretically on statistical grounds to apply to crystals at very low temperatures. Except for metals and other crystals which show electrical conductivity, $a = 0$. For purposes of estimating the small entropy increment between 0 °K and the lowest experimental temperatures in constant-pressure calorimetry it is usually satisfactory to use an equation of the same form for C_P

$$C_P = aT + bT^3 \qquad (7\text{-}12)$$

PROBLEM 7-5 Show that if T' is in the range in which 7-12 is a satisfactory approximation, the entropy increment from 0 °K to T' for a nonmetal will be $\frac{1}{3}bT'^3$. This is evidently equal to $\frac{1}{3}C_P$ where C_P is the observed heat capacity at T'.

::

PROBLEM 7-6 Use the result of Problem 7-5 and the value of C_P shown in Table 7-1 to estimate the entropy increment for AgCl between 0° and 10.5 °K.

::

PROBLEM 7-7 Use one of the graphs of Problem 7-4 together with the answer to Problem 7-6 to calculate ΔS_P, the entropy increment of AgCl between 0° and 298 °K. Compare your result with the value of 22.97 cal deg^{-1} calculated by Eastman and Milner.

::

PROBLEM 7-8 Table 7-2 gives selected values[4] for the heat capacity of silver. One can estimate by using the values for 1° and 5° from Table 7-2 that the values of the constants in 7-12 are $a = 1.4 \times 10^{-4}$ cal deg^{-2} and $b = 4.10 \times 10^{-5}$ cal deg^{-4}. Using these constants calculate values of

Table 7-2   Heat Capacity of Silver

(cal deg^{-1} mole^{-1})

T °K	C_P	T °K	C_P
1	0.000187	30	1.14
2	0.000618	40	2.00
3	0.00154	50	2.78
4	0.00319	75	4.10
5	0.00580	100	4.82
10	0.0480	150	5.49
15	0.162	200	5.80
20	0.400	250	5.97
25	0.747	298.15	6.07

[4] R. Hultgren, R. L. Orr, P. D. Anderson and K. K. Kelley, *Selected Values of Thermodynamic Properties of Metals and Alloys*, John Wiley and Sons, New York, 1963.

C_P at 1, 2, 3, 4, 5, 10, and 15° and show that the calculated values agree very well with those listed through 5° but begin to deviate from the experimental results at 10° and 15°.

::

PROBLEM 7-9 By graphical integration calculate ΔS for heating silver from 0° to 298 °K. Compare your answer with the value of 10.20 cal mole^{-1} deg^{-1} listed by Hultgren *et al.*

::

We also need to calculate the entropy change that chlorine undergoes when its temperature is changed from 0° to 298 °K. We use the data of Giauque and Powell.[5] The values that they give for the transition temperatures and the corresponding ΔH's are

$$\text{M.P.} = 172.1 \text{ °K}, \qquad \Delta H_{fs} = 1531 \text{ cal mole}^{-1}$$

$$\text{B.P.} = 239.1 \text{ °K}, \qquad \Delta H_{v} = 4878 \text{ cal mole}^{-1}$$

By methods which we have discussed (except for the correction to the ideal gas state) they obtain the values shown in Table 7-3.

Although for our present purposes we are interested in the entropy of real chlorine at 298 °C, it is instructive to examine the correction of 0.12 cal deg^{-1} which Giauque and Powell add to put the results on an ideal-gas

Table 7-3   Entropy of Chlorine (1 atm)

(cal deg^{-1} mole^{-1})

0°–15 °K (extrapolation)	0.331
15°–172.12 °K (graphical)	16.573
Fusion (1531/172.12)	8.895
172.12°–239.05 °K (graphical)	5.231
Vaporization (4878/239.05)	20.406
Entropy of actual gas at boiling point	51.44
Correction for gas imperfection	0.12
Entropy of ideal gas at boiling point	51.56

From heat capacity of Cl_2 (g) from boiling point to 298.15 °K[a]

Entropy of real gas at 298.15 °K	53.26[b]
Entropy of ideal gas at 298.15 °K	53.32

[a] Giauque and Powell used 0 °C = 273.10 °K rather than the now-accepted value of 273.15 °K.
[b] Not calculated by Giauque and Powell.

[5] W. F. Giauque and T. M. Powell, *J. Am. Chem. Soc.*, **61**, 1970 (1939).

basis at the boiling point. The actual equation of state of chlorine near its boiling point is given by Eucken and Hoffmann[6] as

$$V = \frac{RT}{P} + 39 - \frac{8.5 \times 10^8}{T^{2.6}} \text{ cm}^3 \text{ mole}^{-1} \tag{7-13}$$

To make the correction Giauque and Powell made use of the equation

$$\left(\frac{\partial S}{\partial P}\right)_T = -\left(\frac{\partial V}{\partial T}\right)_P \tag{7-4}$$

It is probably not immediately obvious how the correction is to be made with the help of 7-4. What we have to do is to find the difference between the entropy increment for isothermally (at the boiling point of chlorine) changing the pressure on an ideal gas and that for changing the pressure on chlorine gas both from some pressure low enough so that Cl_2 is behaving as an ideal gas to 1 atm.

For such an isothermal expansion 7-4 may be rewritten

$$dS = -\left(\frac{\partial V}{\partial T}\right)_P dP \tag{7-14}$$

If at any pressure the differences between the ideal gas and Cl_2 are represented by ΔS and ΔV, 7-14 leads to

$$d\,\Delta S = -\left(\frac{\partial \Delta V}{\partial T}\right)_P dP \tag{7-15}$$

If we use 7-13 as the equation of state for Cl_2 at its boiling point we get

$$\Delta V = \frac{8.5 \times 10^8}{T^{2.6}} - 39 \text{ cm}^3 \text{ mole}^{-1} \tag{7-16}$$

PROBLEM 7-10   By using 7-16 and integrating 7-15 from $P = 0$ to $P = 1$ atm derive the correction to be applied to the entropy of real gaseous chlorine at its normal boiling point to get the value of the entropy that chlorine gas would have at 1 atm and its normal boiling point in the hypothetical ideal gas state. In comparing your answer to this problem with the value of the correction used by Giaque and Powell it may be noted that in their calculation they used Berthelot's equation of state rather than 7-13.

::

We now have the entropy increments for silver, chlorine, and silver chloride for the change in state at 1 atm from 0° to 298 °K, that is, we have values for ΔS_P and ΔS_R of Figure 7-1. To get the value of ΔS at 298 for reaction 2-5 we use the emf data of Gerke.[7] He measured ε and $\partial \varepsilon / \partial T$ for the cell

$$\text{Ag, Ag Cl, HCl } (1M) + \text{Ag Cl, Cl}_2 \text{ (1 atm)} \tag{7-17}$$

[6] Eucken and Hoffman, *Zeit. physik. Chem.*, **B5**, 442 (1929).
[7] Roscoe H. Gerke, *J. Am. Chem. Soc.*, **44**, 1692 (1922).

He found ε at 25 °C = 1.13618 v. For the average value of $\partial\varepsilon/\partial T$ he gives -0.000595 v deg^{-1}. An examination of the data indicates that the latter value is uncertain to at least one percent, and the data are too thin to permit estimation of the variation of $d\varepsilon/dT$ over the 15° to 35 °C range over which the experiments were conducted. From 7-9 we calculate ΔS_{298}:

$$\Delta S_{298} = -23.06 \times 10^3 \times 5.95 \times 10^{-4}$$
$$= -13.7_2 \pm 0.15 \text{ cal deg}^{-1}$$

Then, using the values for ΔS_P and ΔS_R calculated in the way we have indicated and given by the cited authors, and using the above-calculated value for ΔS_{298} we get

$$\Delta S_0 = 26.63 + 10.20 - 22.97 - 13.7_2$$
$$= 0.1 \text{ cal deg}^{-1}$$

This value is zero within the uncertainty of the value used for $d\varepsilon/dT$. Within the limitations of our data, then, we have shown that statement 7-1 is correct with respect to the chlorine–silver–silver chloride reaction. Many other experimental verifications could be cited.[8]

Beyond the foregoing treatment of the Ag–Cl$_2$–AgCl case, very little explanation is required to make it clear how the Third Law provides a widely-applicable means for determining the entropies of elements and compounds at any temperatures of experimental interest, and from these entropies the values for the ΔS's of chemical reactions. Thus, with S_0 taken as zero, S_T for an element or compound can be determined by calorimetry through Equations 4-9 and 4-10 in the manner of the preceding example. Then the ΔS for any reaction will be the difference between the sum of the entropies of the products and the sum of the entropies of the reactants.

$$\Delta S_{\text{Reaction}} = \sum S_{\text{Products}} - \sum S_{\text{Reactants}} \tag{7-18}$$

Entropies of gases can be calculated very accurately by statistical methods from spectroscopic data. The entropy of one substance involved in a chemical reaction can be calculated (without calling in the Third Law) if the entropies of all the other substances involved in the reaction are known and if ΔS for the reaction can be measured by a Second-Law method; but this requires the possibility of carrying out the reaction reversibly. The important field of service of the Third Law is with respect to substances for which the spectroscopic or the Second-Law methods are not available.

[8] See, for example, G. S. Parks and H. M. Huffman, *The Free Energies of Some Organic Compounds*, Reinhold (1932) and Lewis and Randall, *Thermodynamics*, revised by Pitzer and Brewer, McGraw-Hill (1961).

8

Thermodynamic Properties Calculated by Statistical Methods

Guggenheim[1] has proposed that the Third Law be augmented. His proposed statement of the "third principle" is:

By the standard methods of statistical thermodynamics it is possible to derive for certain entropy changes general formulae which cannot be derived from the *zeroth, first or second principles of classical thermodynamics*. In particular one can obtain formulae for entropy changes in highly dispersed systems (i.e., gases), for those in very cold systems (i.e. when $T \to 0$) and for those associated with the mixing of very similar substances (e.g., isotopes).

In the preceding chapter it was acknowledged that there are exceptions to our practical statement of the Third Law. In this chapter we see that the Boltzmann postulate of the relation between entropy and probability provides a reasonable basis for both the law and the exceptions. We also see that methods, which have become exceedingly useful and are very accurate, exist for calculating the thermodynamic properties of ideal gases from spectroscopic data. These methods are based on quantum and statistical mechanics.

Although his statement is less a principle in itself than it is an affirmation of faith in the principles on which the "standard methods" to which it refers are based, Guggenheim's proposal has much merit. One cannot but view with admiration the great power of strictly classical thermodynamics. But modern thermodynamics has made increasing use of developments in the fields of quantum mechanics, statistical mechanics, and the structure of matter, and it is not practical to try to wall off classical from statistical thermodynamics.

---

[1] E. A. Guggenheim, *Thermodynamics, An Advanced Treatment for Chemists and Physicists*, Fifth Ed., North-Holland Publishing Co., Amsterdam, 1967.

Nevertheless, one cannot learn everything at once and most of us cannot learn nearly everything regardless of how long and diligently we work at it. No attempt is made in this book to go very far into the foundations of statistical thermodynamics. Rather, we first consider some ideas about probabilities and distributions in terms of very simple models, and then we proceed to set forth some of the formulas which have been derived from the postulates of statistical thermodynamics (including those of quantum mechanics) which can be used to correlate the thermodynamic properties of a substance with its microstructure and quantized energy states.

THERMODYNAMIC PROBABILITY

Many experiments have firmly established, although still incompletely, the nature of matter as represented by the models of kinetic, atomic, and molecular theory. A rough description of these models is that substances are composed of submicroscopic particles including molecules, atoms, ions, electrons, and nucleii. For chemical purposes we seldom, if ever, have to inquire into the structure of the nucleus. These particulate constituents of substances are, under various constraints, constantly changing position and exchanging energy. The macroscopic properties of a system at a given moment are determined by the state of its substructure at that moment with respect to energy and particle distribution. On the basis of this picture one might expect the macroscopic properties to have different values at different moments of observation.

Experience shows, however, that within limits at least as narrow as the accuracy of the observations many systems have macroscopic properties which do not change from observation to observation. There are two possibilities which would lead to this result. One is that the exchanges of position and energy are occurring *on the average* with equal frequency in opposite directions and that the observation time is very long with respect to this frequency so that the measured value of the macroscopic property represents an average value which will not change detectably from measurement to measurement. The other possibility is that certain kinds of exchanges occur rapidly enough to conform to the situation described above but others which are not occurring with equal frequency on the average in opposite directions are, however, occurring so slowly with respect to the period over which measurements are carried out that they do not produce any detectable change in the observed properties. It is evident that these two possibilities correspond to concepts of equilibrium which have been discussed in preceding chapters.

A particular distribution of particles and energies in a specified system may be called a microstate of the system. The average, macroscopic, or thermodynamic state is determined by the frequency with which the various microstates appear. This frequency may be thought of either in terms of repeated instantaneous observations on the same equilibrium system or of instantaneous observations on many systems which are identical in terms of their macroscopic description. The ordinary *probability* of a particular microstate is defined as the ratio of the number of times that that microstate would appear, if instantaneous observations could be made, to the total number of observations, after a very large number of observations had been made.

Certain ideas about probability, microstates, and macrostates are easily explained in terms of coin-flipping. Suppose that the system consists of 5 coins. The property in terms of which we specify our macrostates and microstates is the ratio of heads to tails. We know, in terms of the discussion in the preceding paragraph, that the macroscopic state corresponds to a value of 1 for this ratio.

What are the microscopic states and their probabilities? Here it is necessary for us to consider two different ways of thinking about the meaning of a distribution. One way is to be concerned not only with the numbers of heads and tails showing after a particular throw but also with which particular coins fall heads and which ones fall tails. Thus we might name the coins A, B, C, D, and E. A possible distribution among heads (H) and tails (T) might then be described as A–H, B–H, C–T, D–T, E–T; another as A–H, B–H, C–H, D–H, E–H. Inasmuch as each coin is assumed to have equal likelihood of falling H or falling T, each of the above arrangements is equally probable.

On the other hand, if we are concerned only with the total number of heads and the total number of tails after a particular throw (the thermodynamic properties of an ideal gas will not depend upon *which* molecules are in each energy state—only upon the total numbers in each state), numbers of the microstates we were considering in the preceding paragraph will be equivalent to each other and will have to be grouped together as a single distribution. Thus, on this basis a single distribution describable as 1H–4T will include all 5 of the microstates represented by the configurations:

     1. A–H, B–T, C–T, D–T, E–T

     2. A–T, B–H, C–T, D–T, E–T

     3. A–T, B–T, C–H, D–T, E–T

     4. A–T, B–T, C–T, D–H, E–T

     5. A–T, B–T, C–T, D–T, E–H

PROBLEM 8-1 On the basis of equal probability of each microstate like those described immediately above show that the relative probabilities of the distributions 5H–0T:4H–1T:3H–2T:2H–3T:1H–4T: 0H–5T are 1:5:10:10:5:1.

A somewhat more elaborate analogy is a set of roulette wheels. Suppose that, although otherwise honest, the wheels are controlled in such a way that they can stop only when the numbers on which the wheels stop add up to some specified and constant sum. This constraint is analogous to the fact that the total energy of an isolated system of particles or of one at equilibrium with its surroundings must remain constant.[2] This analogy helps us to understand the idea of equal *a priori* probability of nondegenerate energy levels. If the sectors of the wheels corresponding to each number are of equal size, there would be equal chance of a wheel stopping on any of the numbers if no over-all energy constraint existed.

Figure 8-1 represents a 9-molecule system (9 roulette wheels). The molecules are identical except with respect to possible differences in their quantum states. Each is assumed to have 3 allowed energy levels: 0, 1, 2. The total energy of the system is arbitrarily taken as 3, i.e., as an average of $\frac{1}{3}$

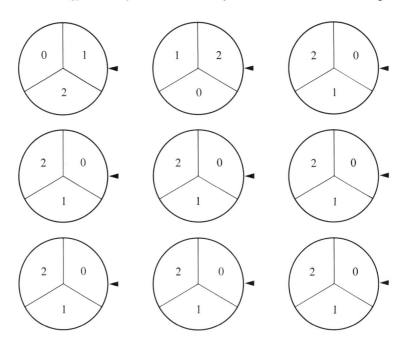

FIGURE 8-1   A 9-molecule (9 roulette wheels) system. The molecules are identical and each has three allowed energy levels—0, 1, 2. The quantum state shown in this figure is one of the 156 allowed states for the case where the total energy of the system is 3 (average of $\frac{1}{3}$ per molecule).

[2] Strictly speaking, the energy of a system in equilibrium with surroundings need not remain constant with microscopic precision. There will be fluctuations, but with macroscopic systems these fluctuations turn out to have a negligible effect. For a short discussion of this problem and its relation to Gibbs' microcanonical and canonical ensembles see Frank C. Andrews, *Equilibrium Statistical Mechanics*, John Wiley and Sons, Inc., 1963.

per molecule. The macroscopic properties of this system depend only upon the total number of molecules which, on the average, are in each of the energy levels. There are only 2 possible distributions. One is with 6 molecules in the 0 state and 3 in the 1 state, the other with 7 molecules in the 0 state and 1 each in the 1 and 2 states. These distributions can be represented as

A
| $n_0 = 6$ |
| $n_1 = 3$ |
| $n_2 = 0$ |

and

| $n_0 = 7$ |
| $n_1 = 1$ |
| $n_2 = 1$ |
B

First we calculate how many different microstates are included in each of the above distributions. We mean by this question exactly the same thing we meant by asking how many different ways there are that 5 coins can fall 3 heads and 2 tails. The total number of permutations of n molecules is $n!$. But permutations among molecules already in the same energy level do not give rise to new microstates. If there are in a particular combination n_0 molecules in the 0 state, n_1 molecules in the 1 state, n_2 is the 2 state, etc., the fraction of the total number of permutations which actually gives rise to a new microstate is $1/n_0! \, n_1! \, n_2! \cdots$ and the formula for the number of different microstates, W, corresponding to a particular distribution will be

$$W = \frac{n!}{n_0! \, n_1! \, n_2! \cdots} \tag{8-1}$$

PROBLEM 8-2   Use 8-1 to show that the values of W (number of microstates) corresponding to the two possible distributions A and B of the 9-molecule system are $W_A = 84$ and $W_B = 72$.

::

PROBLEM 8-3   Show that for a 12-molecule system with allowed energy levels 0, 1, 2 and with the average energy per molecule of $\frac{1}{3}$, there will be three allowed distributions:

A
| $n_0 = 8$ |
| $n_1 = 4$ |
| $n_2 = 0$ |

B
| $n_0 = 9$ |
| $n_1 = 2$ |
| $n_2 = 1$ |

C
| $n_0 = 10$ |
| $n_1 = 0$ |
| $n_2 = 2$ |

and that the values of W will be $W_A = 495$, $W_B = 660$, $W_C = 66$.

::

The thermodynamic (macroscopic) properties of the system do not represent those of any of the contributing distributions such as A, B, and C in the preceding problem but correspond to the average properties of all of the contributing distributions. Suppose that we had a collection (ensemble) of systems such that every system was in a different microstate and every microstate possible under the existing constraints was represented. The total number of systems in this collection would be the sum of the W's for the allowed distributions. The total number of molecules would be the sum of the W's (number of systems) times the number of molecules in each system. The total number of molecules (in the ensemble) in a particular energy level would be the sum, over all distributions, of the products of the number in that level in a particular distribution times the W for that distribution.

We illustrate this way of averaging by reference to the 12-wheel game. The total number of systems required for the particular kind of ensemble we have described would be (see Problem 8-3) $495 + 660 + 66 = 1221$. The total number of molecules (wheels) would be $1221 \times 12 = 14652$. The number of molecules in the 0 energy level would be $8 \times 495 + 9 \times 660 + 10 \times 66 = 10560$. If we call n the total number of molecules, then in our ensemble $n_0/n = 10560/14652 = 0.720$. We can interpret this ratio as the fraction of molecules we would find in the zero level if we had a very large number of systems each constrained to contain 12 molecules and to have a total energy of 4 units. We can also interpret the ratio as the one we would get by averaging the results of many many observations on a single system of the kind described.

By making analogous calculations for the 1 and 2 energy levels we get the following properties for the average system:

$$\left.\begin{aligned} \frac{n_0}{n} &= \frac{10{,}560}{14{,}652} = 0.720 \\[6pt] \frac{n_1}{n} &= \frac{3300}{14{,}652} = 0.225 \\[6pt] \frac{n_2}{n} &= \frac{792}{14{,}652} = 0.054 \end{aligned}\right\} \begin{array}{l} \text{Average} \\ \text{system} \end{array}$$

It is interesting to compare the above ratios with the ratios corresponding to the most probable system B. These are

$$\left.\begin{aligned} \frac{n_0}{n} &= \frac{9}{12} = 0.750 \\[6pt] \frac{n_1}{n} &= \frac{2}{12} = 0.167 \\[6pt] \frac{n_2}{n} &= \frac{1}{12} = 0.083 \end{aligned}\right\} \begin{array}{l} \text{Most probable} \\ \text{system} \end{array}$$

PROBLEM 8-4 Show that the requirement that the average energy be $\frac{1}{3}$ per molecule is met by the distribution shown above for the average system.

PROBLEM 8-5 With a 15-molecule system, 0, 1, 2 energy levels for each molecule, and specified average energy of $\frac{1}{3}$ per molecule, the allowed combinations would be

	A		B		C
$n_0 = 10$		$n_0 = 11$		$n_0 = 12$	
$n_1 = 5$		$n_1 = 3$		$n_1 = 1$	
$n_2 = 0$		$n_2 = 1$		$n_2 = 2$	

Use 8-1 to find that $W_A = 3003$, $W_B = 5460$, $W_C = 1365$.

::

PROBLEM 8-6 By the procedure described and carried out on the 12-molecule case calculate for the 15-molecule case the relative numbers of molecules in the several energy levels for the average state of the system and also for the most probable state.
Answer:

$$
\left.
\begin{aligned}
\frac{n_0}{n} &= 0.721 \\[2mm]
\frac{n_1}{n} &= 0.224 \\[2mm]
\frac{n_2}{n} &= 0.055
\end{aligned}
\right\} \begin{array}{l}\text{Average}\\\text{state}\end{array}
$$

$$
\left.
\begin{aligned}
\frac{n_0}{n} &= 0.734 \\[2mm]
\frac{n_1}{n} &= 0.200 \\[2mm]
\frac{n_2}{n} &= 0.067
\end{aligned}
\right\} \begin{array}{l}\text{Most probable}\\\text{state}\end{array}
$$

::

PROBLEM 8-7 Calculate the distributions in the average state and in the most probable state for the 9-wheel game (as was done in Problem 8-6 for the 15-wheel system).

::

A comparison of the average distributions with the most probable distributions as calculated for the 9-, 12-, and 15-molecule systems suggests a correct and very important conclusion. It is that as the number of molecules in the system increases, the average and the most probable distributions become more nearly the same. In fact, in our examples the two distributions have already become so similar with only 15 molecules that we feel entirely comfortable with the extrapolation that the difference between the average distribution and the most probable distribution in a system having of the order of Avogadro's number of molecules would be utterly negligible. Another correct conclusion probably suggested by a comparison of the results on the 9-, 12-, and 15-molecule systems which have been calculated is that for systems having the same average energy per molecule, the relative distributions of molecules among the several energy levels approach constant values not dependent on the number of molecules as that number becomes larger.

PROBLEM 8-8   Show for the 9-wheel case, if the average energy is made $\frac{2}{3}$ per wheel instead of $\frac{1}{3}$, the value of W for the most probable distribution is much larger.

Answer: For the average energy of $\frac{1}{3}$ we found for the most probable distribution $W = 84$; with the average energy of $\frac{2}{3}$ you should find for the most probable distribution $W = 756$.

∷

With the limited number of energy levels we have represented on the roulette wheels, the result of Problem 8-8 cannot be extrapolated to the conclusion that a continued increase in the average energy would lead to a continued increase in W. In fact, with these wheels as described the maximum possible average energy would be 2, and for a system with this average energy the value of W would be 1. However, with systems of molecules the number of allowed energy levels is practically limitless and calculated values of W will be found to increase continuously as the average energy is made greater and greater.

Boltzmann[3] gave the name *thermodynamic probability* (Wahrscheinlikeit) to W and postulated the relation between the entropy of a system in a particular state and the thermodynamic probability of that state:

$$S = k \ln W \qquad (8-2)$$

Before exploring the implications of this postulate and testing it out quantitatively in a few cases that can readily be calculated, we consider its plausibility in the light of what we know about entropy in terms of classical thermodynamics and what we know about W from the preceding sections of this chapter. We know on the one hand that the entropy of isolated systems tends to increase to some maximum value corresponding to the equilibrium state; we know on the other hand that systems which are shaking around tend to come to the most probable allowed distribution, a distribution which is characterized by a maximum value for W.

We know also that entropy is an extensive property and that the total entropy of two macroscopically identical systems is just twice that of one of them. But W does not behave that way. If there are W possible arrangements of system 1 and W possible arrangements of system 2, then there

[3] Ludwig Boltzmann, *Vorlesung uber Gastheorie*, Johann Ambrosius Barth Verlag, Leipzig, 1912.

are W^2 possible arrangements of a double system composed of system 1 plus system 2. The logarithmic relation is just what is needed for $k \ln W^2 = 2k \ln W$.

Too, the effect on the entropy of temperature changes is at least qualitatively consistent. Problem 8-8 suggests the correct conclusion that W increases as the energy in a system increases. But at constant volume, for example, this energy increment is

$$dE = C_V \, dT \qquad (1\text{-}40)$$

So W increases when the temperature of a system is increased. But we have also for constant volume heating

$$dS = C_V \frac{dT}{T} \qquad (4\text{-}10)$$

Thus, both the Boltzmann equation, 8-2, and our Second Law definition of the entropy increment indicate an increase in entropy as the temperature of a system is raised.

BOLTZMANN STATISTICS AND
THE PARTITION FUNCTION

The Boltzmann distribution law for ideal gases can be readily derived on the basis of the ideas that have been discussed above and have been illustrated by the foregoing problems. Consider an isolated system of n molecules. Two conditions that must exist regardless of the distribution of energy among the various energy levels of the various molecules are that the total number of molecules remains constant and that the total energy remains constant. Letting n_i represent the number of molecules in the ith energy level, the energy of which is ε_i, the total number of molecules and the total energy are:

$$n = \sum n_i \qquad (8\text{-}3)$$

and

$$E = \sum n_i \varepsilon_i \qquad (8\text{-}4)$$

The conditions of constant number and constant energy can then be stated by indicating that the variations must be zero.

$$\delta n = \delta \sum n_i = \sum \delta n_i = 0 \qquad (8\text{-}5)$$

and

$$\delta E = \delta \sum n_i \varepsilon_i = \sum \varepsilon_f \, \delta n_i = 0 \qquad (8\text{-}6)$$

The Boltzmann distribution applies to a system at equilibrium. For an equilibrium system of a large number of molecules our examples have suggested and we have asserted that the average distribution is identical for any thermodynamic purposes to the most probable distribution. The most probable distribution is the one for which W is a maximum. At points where functions have maximum and minimum values, the variations of these functions are zero. For example, the variation in y with small changes in x, dy/dx, is zero at points where y has maximum or minimum values. We can then write down our third condition (the equilibrium distribution is the one for which W is maximum) as

$$\delta W = 0 \qquad (8\text{-}7)$$

Let us recast 8-7 into a logarithmic form. If the variation in W is zero then the variation in $\ln W$ must also be zero, hence the condition expressed by 8-7 is also expressed by

$$\delta \ln W = 0 \qquad (8\text{-}8)$$

We can see in the light of 8-2 that by 8-8 we are characterizing the equilibrium distribution as the one of maximum entropy.

Before going further with the derivation of the Boltzmann distribution we need to consider the quantum weight or degree of degeneracy of energy levels. Suppose that several quantum states of a molecule differ only in their magnetic quantum number. Each of these quantum states should have equal *a priori* probability with respect to each other and to all other quantum states of the molecule. But in the absence of a magnetic field each of these states will have the same energy. This fact is the basis of the use of the term *degenerate* energy level. Several quantum states which in the presence of a magnetic field represent different energy levels have degenerated into a single energy level in the absence of the field. The degree of degeneracy of the ith energy level is the number of quantum states each having the energy ε_i. This degree of degeneracy is given the symbol g_i.

PROBLEM 8-9   The effect of degenerate energy levels on the distribution of energy can be illustrated by roulette wheels of the kind shown in Figure 8-2. This wheel represents a molecule for which $g_0 = 1$, $g_1 = 2$, and $g_2 = 3$. Consider 9 wheels of this kind for which the average energy is $\frac{1}{3}$ per wheel. In the 9-wheel case considered above there was no degeneracy. In that case for an average energy of $\frac{1}{3}$ per wheel the two possible distributions were

$$
A \quad
\begin{array}{|c|}
\hline
n_0 = 6 \\
n_1 = 3 \\
n_2 = 0 \\
\hline
\end{array}
\quad \text{and} \quad
\begin{array}{|c|}
\hline
n_0 = 7 \\
n_1 = 1 \\
n_2 = 1 \\
\hline
\end{array}
\quad B
$$

and the values of W were $W_A = 84$, $W_B = 72$.

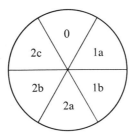

FIGURE 8-2   Roulette wheel representing a molecule in which $g_0 = 1$, $g_1 = 2$, and $g_2 = 3$.

In the case of the 9 wheels of the kind shown in Figure 8-2 there are additional distributions as shown below:

A

$n_0 = 6$	$n_0 = 6$	$n_0 = 6$	$n_0 = 6$
$n_{1a} = 3$	$n_{1b} = 3$	$n_{1a} = 2$	$n_{1a} = 1$
$n_{1b} = 0$	$n_{1a} = 0$	$n_{1b} = 1$	$n_{1a} = 2$
$n_2 = 0$	$n_2 = 0$	$n_2 = 0$	$n_2 = 0$

B

$n_0 = 7$
$n_{1a} = 1$
$n_{2a} = 1$
$n_{1b} = 0$
$n_{2b} = 0$
$n_{2c} = 0$

etc.

All of the distributions included in A will represent a single distribution as far as their contributions to the macroscopic (thermodynamic) properties of the system are concerned, and all the B distributions, likewise, can be counted as identical contributions to the thermodynamic state. By calculating through 8-1 the total numbers of microstates in A and B show that $W_A = 672$, $W_B = 432$. Show that these results are in accord with 8-9.

::

In order to take degeneracy into account 8-1 must be modified. The correct equation for the value of W for a particular distribution (in the sense that A and B in the above problem are each considered single distributions) is

$$W = n! \prod \frac{g_i^{n_i}}{n_i!} \qquad (8\text{-}9)$$

The corresponding equation for $\ln W$ is

$$\ln W = \ln n! + \sum (n_i \ln g_i - \ln n_i!) \tag{8-10}$$

We can now proceed with the derivation of Boltzmann's distribution law by using 8-10 with 8-8. First we rewrite 8-10 in terms of a form of Stirling's approximation which is entirely satisfactory for large values of n. It is

$$\ln n! = n \ln n - n \tag{8-11}$$

Using 8-11 we get from 8-10

$$\ln W = \ln n! + \sum (n_i \ln g_i - n_i \ln n_f + n_i) \tag{8-12}$$

PROBLEM 8-10   Remembering that n is a constant and also that the g's are constants, show by differentiation that 8-12 leads to

$$\delta \ln W = \sum \ln \frac{g_i}{n_i} \delta n_i \tag{8-13}$$

::

Using 8-13 in 8-8 we can write

$$\sum (\ln g_i - \ln n_i) \delta n_i = 0 \tag{8-14}$$

The three conditions for the equilibrium distribution, then, are represented by

$$\sum \delta n_i = 0 \tag{8-5}$$

$$\sum \varepsilon_i \delta n_i = 0 \tag{8-6}$$

and 8-14.

PROBLEM 8-11   Using Lagrange's method of undetermined multipliers one can obtain simultaneous solutions to these equations. Multiply 8-5 by α and 8-6 by β and show that

$$n_i = g_i e^{(\alpha + \beta \varepsilon_i)} \tag{8-15}$$

(For an elementary discussion of the use of Lagrangian multipliers see, for example, Myron Tribus, *Thermostatics and Thermodynamics*, pp. 71–74; D. Van Nostrand & Co., 1961.)

::

We do not need to identify α explicitly because we are always interested in ratios. For example

$$\frac{n_i}{n_0} = \frac{g_i}{g_0} e^{\beta(\varepsilon_i - \varepsilon_0)} \tag{8-16}$$

There are various ways of identifying β. One is to take as an implicit statistical thermodynamic definition of T that

$$\beta = \frac{-1}{kT} \tag{8-17}$$

and then by deriving an expression for the translational energy of an ideal gas to show that the T in 8-17 is identical with the T in $PV = nRT$ which we have shown in terms of classical thermodynamics to be the same as the thermodynamic temperature. In any case we rewrite 8-16

$$\frac{n_i}{n_0} = \frac{g_i}{g_0} e^{-(\varepsilon_i - \varepsilon_0)/kT} \tag{8-18}$$

This equation we can take as the basis of Boltzmann statistics. For our purposes we do not investigate Bose-Einstein and Fermi-Dirac statistics.

We saw in our examples with the roulette wheels that from the average energy per molecule and a knowledge of the allowed energy levels, the energy distribution could be determined (if the number of molecules was large, the distribution did not depend on the number). The expression for the average energy is

$$\bar{\varepsilon} = \frac{\sum n_i \varepsilon_i}{\sum n_i} \tag{8-19}$$

In terms of 8-15 and 8-17 this becomes

$$\bar{\varepsilon} = \frac{\sum g_i \varepsilon_i e^{-\varepsilon_i/kT}}{\sum g_i e^{-\varepsilon_i/kT}} \tag{8-20}$$

The denominator of 8-20 is most important in statistical thermodynamics. It is usually called the *partition function* (sometimes it is called *sum-over-states*) and is given the symbol Q.

$$Q \equiv \sum g_i e^{-\varepsilon_i/kT} \qquad \begin{bmatrix} \text{The partition} \\ \text{function} \end{bmatrix} \tag{8-21}$$

Q is a measure of the extent of energy distribution among the various energy levels of the system.

> *PROBLEM 8-12*   By differentiation of 8-21 show that 8-20 gives
>
> $$\bar{\varepsilon} = kT^2 \left(\frac{\partial \ln Q}{\partial T} \right) \tag{8-22}$$
>
> ■■

STATISTICAL CALCULATION OF THERMODYNAMIC PROPERTIES OF IDEAL GASES

In Equation 8-22 we have the prototype of the use of statistical methods in the calculation of thermodynamic properties of ideal gases. Thus, for the molar energy of an ideal gas we can write[4]

$$E - E_0 = RT^2 \left(\frac{\partial \ln Q}{\partial T} \right) \tag{8-23}$$

---

[4] Absolute values of the energy of molecules or systems of molecules are not meaningful. Thus, the ε's in the partition function can only be evaluated relatively to each other. Because of these facts, we cannot evaluate $\bar{\varepsilon}$ or E except on some arbitrary basis but we can evaluate the difference between the energy of the system, E_0, at 0 °K and its energy, E, at some other temperature.

The data needed to evaluate $E - E_0$ are the data needed to evaluate Q, or more specifically in this case its variation with T. These data are the values of the g_i's and the ε_i's.

Inasmuch as the internal energy levels of molecules of an ideal gas are not affected by the translational energy distribution, it is possible and very convenient to treat the translational contributions to the thermodynamic properties separately from the contributions of the internal energy distributions. Without doing the statistical derivations we here just set down the expressions for the translational contributions which are valid at all but extremely low temperatures. We write here only expressions for E, C_V, and S; the other functions can be readily calculated from these and the easily measured macroscopic properties P and V.

$$(E - E_0)_{tr} = \tfrac{3}{2}RT \tag{8-24}$$

$$C_{V_{tr}} = \tfrac{3}{2}R \tag{8-25}$$

$$S_{tr} = R(\tfrac{5}{2}\ln T + \tfrac{3}{2}\ln M - \ln P - 1.1644) \tag{8-26}$$

8-26 is the Sackur–Tetrode equation. In it M is the molecular weight of the gas, T is in °K and P is in atm.

PROBLEM 8-13   Calculate the molar translational entropy of HCl in the hypothetical ideal gas state at 1 atm, 25 °C.

::

The internal energy distribution in an ideal gas is not dependent on the pressure or volume. To get expressions in terms of Q for the internal energy contributions to the thermodynamic properties we begin with the heat capacity, C_{int}. Using 8-23 we have

$$C_{int} = \frac{dE_{int}}{dT} = R\frac{d}{dT}\left(T^2\frac{d\ln Q_{int}}{dT}\right) \tag{8-27}$$

For convenience we sometimes omit the subscript *int* or other subscripts on the Q's but it should be understood that the Q's in equations such as 8-27 refer to the partition functions of the kinds of energies whose contribution is being calculated whether they include all the internal energies as in 8-27 or whether some particular kind such as vibrational energies.

PROBLEM 8-14   By differentiation show that

$$C_{int} = R\left(T^2\frac{d^2\ln Q}{dT^2} + 2T\frac{d\ln Q}{dT}\right) \tag{8-28}$$

::

PROBLEM 8-15 Show that 8-28 is equivalent to

$$C_{int} = RT^2 \frac{1}{Q} \frac{d^2Q}{dT^2} - RT^2 \frac{1}{Q^2} \left(\frac{dQ}{dT}\right)^2 + 2RT \frac{1}{Q} \frac{dQ}{dT} \qquad (8\text{-}29)$$

■■

PROBLEM 8-16 By differentiating 8-21 get expressions for dQ/dT and d^2Q/dT^2 and show that 8-29 can be written as

$$C_{int} = R\left[\frac{1}{Q} \sum g_i \left(\frac{\varepsilon_i}{kT}\right)^2 e^{-\varepsilon_i/kt} - \frac{T^2}{Q^2}\left(\frac{dQ}{dT}\right)^2\right] \qquad (8\text{-}30)$$

■■

PROBLEM 8-17 By defining Q' and Q'' as

$$Q' = \sum g_i \frac{\varepsilon_i}{kT} e^{-\varepsilon_i/kT} \qquad (8\text{-}31)$$

$$Q'' = \sum g_i \left(\frac{\varepsilon_i}{kT}\right)^2 e^{-\varepsilon_i/kT} \qquad (8\text{-}32)$$

show that C_{int} can be written as

$$C_{int} = R\left[\frac{Q''}{Q} - \left(\frac{Q'}{Q}\right)^2\right] \qquad (8\text{-}33)$$

■■

For cases in which the partition function can be expressed to satisfactory approximation in a form subject to convenient differentiation with respect to temperature, expressions like 8-28 for the thermodynamic properties are useful. For cases in which direct summations must be used, forms like 8-33 are convenient.

PROBLEM 8-18 Show that

$$(E - E_0)_{int} = RT \frac{Q'}{Q} \qquad (8\text{-}34)$$

■■

PROBLEM 8-19 For the contribution of the internal energy distributions to the entropy we can write

$$(S - S_0)_{int} = \int_0^T \frac{C_{int}}{T} \, dT \qquad (8\text{-}35)$$

By using 8-28 for the heat capacity and integrating by parts show that

$$(S - S_0)_{int} = R\left[\ln Q + T \frac{d \ln Q}{dT}\right]_0^T \qquad (8\text{-}36)$$

■■

To put the lower limit into the right-hand member of 8-36 we note that when $T = 0$,

$$Q = g_0 e^{-\varepsilon_0/kT}$$

Then at $T = 0$

$$\ln Q = \ln g_0 - \varepsilon_0/kT$$

and

$$T \frac{d \ln Q}{dT} = \varepsilon_0/kT$$

So the value of the integral at the lower limit is just $R \ln g_0$. If now we take Q to represent the partition function at T, the value of the integral at the upper limit will be just what is in the brackets in 8-36 and we have

$$(S - S_0)_{int} = R\left[\ln Q + T \frac{d \ln Q}{dT} - \ln g_0\right] \qquad (8\text{-}37)$$

From 8-2 and 8-9 and making use of the fact that at $T = 0$, $n = n_0$, we can conclude that

$$W_0 = g_0{}^n \qquad (8\text{-}38)$$

and hence that for a mole of gas

$$S_0 = k \ln W_0 = nk \ln g_0 = R \ln g_0 \qquad (8\text{-}39)$$

Combining 8-37 and 8-39 gives

$$S_{int} = R\left(\ln Q + T \frac{d \ln Q}{dT}\right) \qquad (8\text{-}40)$$

To complete our set of expressions for the internal contributions to the thermodynamic properties in terms of the partition function, we note first that the PV term which relates E and H and relates A and G is taken into account in the translational contributions. There is no difference between E_{int} and H_{int} or between A_{int} and G_{int}. We also can see that for the same, reason that we did not write an expression for E_{int} but wrote it for $(E - E_0)_{int}$ we have to write a similar kind of expression for the free energy. Recalling the definition of G we write

$$G - E_0 = H - TS - E_0 \qquad (8\text{-}41)$$

In the light of the above comments we can write equally correctly

$$(G - H_0)_{int} = (E - E_0)_{int} - TS_{int} \qquad (8\text{-}42)$$

Combining 8-42 with 8-23 and 8-40 gives

$$(G - H_0)_{int} = -RT \ln Q \qquad (8\text{-}43)$$

or

$$\left(\frac{G - H_0}{T}\right)_{int} = -R \ln Q \qquad (8\text{-}44)$$

The expression involving A will be the same. We can write

$$\left(\frac{G - E_0}{T}\right)_{int} = -R \ln Q \tag{8-45}$$

$$\left(\frac{A - E_0}{T}\right)_{int} = -R \ln Q \tag{8-46}$$

$$\left(\frac{A - H_0}{T}\right)_{int} = -R \ln Q \tag{8-47}$$

The convenient cases to treat are those in which the distributions of internal energy corresponding to rotational, vibrational, and electronic energy levels can be considered independent of each other. Fortunately, for many systems this can be done with satisfactory approximation to the actual situation. Moreover, in many cases only the rotational and vibrational distributions make significant contributions to be added to the translational ones. We consider only such cases.

The problem of evaluating the partition function Q is primarily an experimental one of determining the energy levels. The amount of experimental data needed, however, is greatly reduced by the results of quantum mechanics which give relationships between the values of successive energy levels. We consider the case of an unsymmetrical diatomic molecule such as HCl as an example.[5] For the rotational energy levels of such a molecule the following relations hold.

$$\varepsilon_J = J(J + 1)\frac{h^2}{8\pi^2 I} \tag{8-48}$$

$$g_J = 2J + 1 \tag{8-49}$$

In these equations

J is the rotational quantum number

h is Planck's constant

I is the moment of inertia of the molecule

g_J is the degeneracy of the Jth level

J takes all integral values, i.e., $0, 1, 2 \ldots$.

[5] See the paper by W. F. Giauque, "The Calculation of Free Energy From Spectroscopic Data," *J. Am. Chem. Soc.*, **52**, 4808 (1930). See also W. F. Giauque and Roy Overstreet, "The Hydrogen, Chlorine, Hydrogen Chloride Equilibrium at High Temperatures," *J. Am. Chem. Soc.*, **54**, 1731 (1932), for a very readable and clear exposition of the application of spectroscopic data to the calculation of thermodynamic functions and in particular for a discussion of refinements that can be made in the calculations of rotational contributions to take account the facts that (1) the molecules are not rigid, (2) the substitution of an integral for a summation introduces some error, and (3) the gas may be a mixture of isotopes. Finally, as an example of a number of good general textbooks on statistical mechanics the book by J. E. and M. G. Mayer (John Wiley and Sons, Inc., 1940) is cited. Among other contributions it offers to our present topic is a discussion of the procedure of basing the rotational and vibrational energy level scale on an assignment of zero to the ground state.

In general it is impossible completely to extricate the contributions of the different kinds of energy distributions to the thermodynamic properties from one another. For example, the moment of inertia of a molecule and hence the rotational contributions will depend to some extent on the vibrational state. Moreover, even apart from the variation in I with the vibrational state, there will be an increase in I with an increase in J because of the stretching of the molecule by centrifugal force. Better but more complicated approximations are available than the ones we use here, but even the simple ones we use give in many cases quite satisfactory results, as we see in the example of the HCl molecule. One simple approximation we use is that the molecule can be treated as a rigid rotator having a value of I which can be assigned on the basis of measurements of a few absorption peaks in the rotational-vibrational spectrum.

To simplify the notation we define y

$$y = \frac{h^2}{8\pi^2 IkT} \tag{8-50}$$

The rotational partition function of a linear, unsymmetrical molecule is then, from 8-21,[6]

$$Q_{rot} = \sum (2J + 1)e^{-J(J+1)y} \tag{8-51}$$

If the values of ε_J are relatively closely spaced (compared to the average rotational energy per molecule) the sum in 8-51 can be well approximated by the integral

$$Q_{rot} = \int_0^\infty (2J + 1)e^{-J(J+1)y}\,dJ \tag{8-52}$$

Figure 8-3 indicates the nature of this approximation.

PROBLEM 8-20   By integrating 8-52 show that in cases in which it is a satisfactory expression for Q_{rot},

$$Q_{rot} = \frac{1}{y} \tag{8-53}$$

■■

Usually I, and hence y, is evaluated from spectroscopic data. Hence it is useful to have an expression for Q more directly related to such data.

[6] For this case J takes all integral values. For a symmetrical linear molecule J has only even-numbered or odd-numbered values. This fact will mean that the integral written below (8-52) for the rotational partition function must be divided by 2 (the *symmetry number*) for a symmetrical, linear rotator. The treatment of nonlinear molecules is more complicated in that moments of inertia corresponding to three axes of rotation must be considered and that symmetry numbers other than 1 or 2 may have to be used.

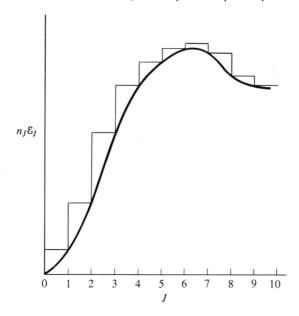

FIGURE 8-3   The relation between Equations 8-51 and 8-52. The area under the step-like line is $\Sigma n_J \varepsilon_J$, the area under the curve is $\int n_J \varepsilon_J dJ$.

PROBLEM 8-21   First, by a comparison of 8-48 and 8-50 we can see that

$$\varepsilon_J = J(J + 1)ykT \qquad (8\text{-}54)$$

Next, show that

$$\varepsilon_{(J+1)} - \varepsilon_J = 2(J + 1)ykT \qquad (8\text{-}55)$$

in which J is the rotational quantum number of the lower state.

∷

PROBLEM 8-22   We take as the fundamental rotational frequency ν_0 that frequency corresponding to the transition between the states corresponding to $J = 0$ and $J = 1$. Noting that

$$\varepsilon_1 - \varepsilon_0 = h\nu_0 \qquad (8\text{-}56)$$

show that

$$y = \frac{h\nu_0}{2kT} \qquad (8\text{-}57)$$

and hence that

$$Q_{rot} = \frac{2kT}{h\nu_0} \qquad (8\text{-}58)$$

∷

We turn now to vibrational contributions to the thermodynamic properties. Again, we take as the example the diatomic molecule. This molecule can have only a single stretching mode of vibration. In passing, however, it should be pointed out that the treatment of more complicated molecules in which there are various kinds of distortional oscillation possible is scarcely more complicated than for the simple diatomic case. For the polyatomic molecule one can calculate separately, by the same formulas we develop for the diatomic oscillator, the contributions corresponding to each characteristic frequency.

The simplest approximation will be made, namely that the molecule vibrates harmonically. Methods are available for correcting for anharmonicity. The quantum mechanical expression for the energy of the vibrational levels of a harmonic oscillator is usually given as

$$\varepsilon_v = h\nu(v + \tfrac{1}{2}) \tag{8-59}$$

ε_v is the vibrational energy

h is Planck's constant

ν is the vibrational frequency

v is the vibrational quantum number which has all integral
values, 0, 1, 2 . . .

The vibrational levels are nondegenerate, i.e., $g_v = 1$.

Although 8-59 indicates a ground level energy of $\tfrac{1}{2}h\nu$, the Boltzmann distribution 8-18 depends only upon differences in energy between levels. In any case, as has been pointed out earlier, the absolute value of the energy of a system in a defined state is indeterminate and assigned values are based on an arbitrary choice of the state to which zero energy is ascribed. In fact, the value given by 8-59 is itself based on such an arbitrary assignment. We are at liberty, therefore, to choose a different reference state. If we choose the zero level $\tfrac{1}{2}h\nu$ higher than that chosen in the quantum mechanical derivation of 8-59 we get

$$\varepsilon_v = h\nu v \tag{8-60}$$

and we thereby get a simpler, but equally correct and useful, vibrational partition function than we would get from 8-59.

PROBLEM 8-23 From the definition of the partition function (8-21) and 8-60 show that the vibrational partition function is

$$Q_{vib} = \sum_v e^{-h\nu v/kT} \tag{8-61}$$

∷

PROBLEM 8-24   Noting that v takes all integral values, 0, 1, 2, . . . , prove by long division that the sum in 8-61 is equal to

$$\frac{1}{1 - e^{-hv/kT}}$$

and hence that

$$Q_{vib} = (1 - e^{-hv/kT})^{-1} \qquad (8\text{-}62)$$

::

PROBLEM 8-25   For cases in which 8-58 is a satisfactory approximation show that

$$(E - E_0)_{rot} = RT \qquad (8\text{-}63)$$

$$C_{rot} = R \qquad (8\text{-}64)$$

$$S_{rot} = R\left(1 + \ln\frac{2kT}{hv_0}\right) \qquad (8\text{-}65)$$

$$\left(\frac{G - H_0}{T}\right)_{rot} = -R\ln\frac{2kT}{hv_0} \qquad (8\text{-}66)$$

::

PROBLEM 8-26   Show that for the harmonic oscillator

$$(E - E_0)_{vib} = RT\frac{hv}{kT}(e^{hv/kT} - 1)^{-1} \qquad (8\text{-}67)$$

$$C_{vib} = R\left(\frac{hv}{kT}\right)^2 e^{hv/kT}(e^{hv/kT} - 1)^{-2} \qquad (8\text{-}68)$$

$$S_{vib} = R\left[\frac{hv}{kT}(e^{hv/kT} - 1)^{-1} - \ln(1 - e^{-hv/kT})\right] \qquad (8\text{-}69)$$

$$\left(\frac{G - H_0}{T}\right)_{vib} = R\ln(1 - e^{-hv/kT}) \qquad (8\text{-}70)$$

::

For many molecules the fraction of the energy in excited electronic states is negligibly small at ordinary temperatures and can, to a good approximation, be ignored in the statistical calculation of thermodynamic properties. In cases and at temperatures for which this is not a good approximation the calculation is considerably complicated not only by the necessity of summing the electronic distributions but also by the necessity of knowing the rotational and vibrational frequencies in the excited electronic states. There are also other factors such as internal rotation which complicate some cases.

Also, before showing how a treatment which is uncomplicated either experimentally or mathematically can give good results in simple cases, it is pointed out that the analysis of the spectra of polyatomic molecules into an appropriate set of frequencies to use in the expressions for the rotational and vibrational partition functions may be difficult. It is a job for a specialist, the spectroscopist.

Figure 8-4 shows a portion of the vibrational-rotational spectrum of HCl.[7] Here the relative absorption is plotted against the frequency expressed as wave numbers (cm^{-1}).[8] For the analysis of this spectrum only

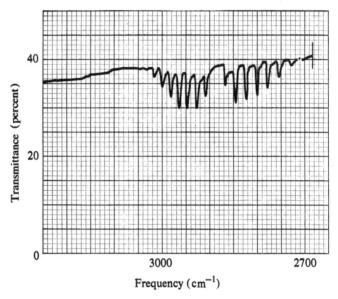

FIGURE 8-4   A portion of the fundamental rotational-vibrational spectrum of HCl.

one further quantum mechanical rule is required. That is that for allowed transitions between vibrational states for this type of molecule $\Delta v = \pm 1$ and $\Delta J = \pm 1$.

Because ΔJ cannot be zero there will be no absorption peak corresponding to the energy of a vibrational transition. Rather, the absorption peaks will correspond to the energy differences between adjacent vibrational levels either increased or decreased by the amount of the rotational

[7] This spectrogram was made as a part of an undergraduate physical chemistry experiment by Dr. Fredrick L. Minn and his students at The George Washington University. A medium-resolution IR spectrophotometer was used. The sample was obtained by bubbling nitrogen through concentrated aqueous HCl at room temperature. The water vapor present did not interfere.

[8] The wave number is the reciprocal of the wavelength and is, therefore, the number of waves per unit length, the length usually being measured in centimeters. The wave number multiplied by the velocity of light is the frequency. Spectroscopic wavelengths can be measured with very high precision.

transition energy. Inasmuch as more of the molecules will be in the vibrational ground state than in any higher vibrational state—and by far more in the present example—the strongest features of the absorption spectrum will be for vibrational transitions between the ground and the first excited vibrational level. Such transitions are responsible for the absorption peaks shown in Figure 8-4. On the other hand, with HCl and heavier molecules at room temperature a number of the lower rotational states are sufficiently populated so that transitions such as $v = 0 \; J = 0 \rightarrow v = 1 \; J = 1$, $v = 0 \; J = 1 \rightarrow v = 1 \; J = 0$, $v = 0 \; J = 1 \rightarrow v = 1 \; J = 2$, $v = 0 \; J = 2 \rightarrow v = 1 \; J = 3$, and so on are all sufficiently frequent to produce prominent peaks in the absorption spectrum.

PROBLEM 8-27   From 8-55 and 8-57 show that for the diatomic rigid rotator-harmonic vibrator one would expect equally spaced absorption peaks one series representing the $v = 0 \rightarrow v = 1$ transition energy *plus* (the **R** branch) successively increasing amounts of rotational transition energy and the other series (the **P** branch) also representing the $v = 0 \rightarrow v = 1$ transition but *minus* successively increasing amounts of rotational energy. That is, show that the first peak in the **R** branch represents the transition $v = 0 \rightarrow v = 1$, $J = 0 \rightarrow J = 1$; that the second peak of this series is $v = 0 \rightarrow v = 1$, $J = 1 \rightarrow J = 2$; etc., and show that these peaks should be equally spaced on an energy or a frequency scale. Show that the spacing is just $h\nu_{0rot}$ on an energy scale or just ν_{0rot} on a frequency scale.

::

PROBLEM 8-28   By estimating the position of the missing peak (the frequency that would correspond to the forbidden transition $v = 0 \; J = a \rightarrow v = 1 \; J = a$ get a value for the fundamental vibration frequency. By estimating the distance between the third peak to the left of the missing peak and the third peak to the right of the missing peak and dividing the result by 6 get a value for ν_{0rot}.

::

Before going further with our calculations, several observations might be made. One is that the peaks are not equally spaced, showing that our simple model does not represent the facts very well. Another is that by taking an average spacing in the region of high absorption we are probably compensating for some of the discrepancy between the model and the actual gas.

The third observation is of a different kind. It is, first, that the population of the ground rotational state will be greater than that of a first excited rotational state, that of a first greater than that of a second excited state, etc. Yet the graph shows that the greatest absorption does not occur for the transitions $J = 0 \rightarrow J = 1$ and $J = 1 \rightarrow J = 0$. The main reason for this is given by 8-49. The total population in 6 degenerate states corresponding to

$J = 2$ may be greater, in spite of the unfavorable Boltzmann factor $(e^{-\Delta\varepsilon_J/kT})$ than in the 2 degenerate $J = 1$ states.

PROBLEM 8-29   Ignore the fact that HCl is isotopically impure but use its chemical molecular weight. On the basis of the theory and formulas that have been developed and using the frequencies estimated in Problem 8-28 calculate the translational, rotational and vibrational contributions to the three thermodynamic properties of HCl in the hypothetical ideal gas state at 1 atm: $H_{298} - H_0$, S_{298}, and C_{P298}. Add these contributions to get the total values of these properties. Probably surprisingly good agreement (in view of our seemingly rather crude measurements and approximations)

Table 8-1   Computer Treatment of a Spectrogram

(see Figure 8-4)

TEST RUN FOR HCL SPECTRUM

FREQUENCY (CM-1)	FIRST DIFFERENCE (CM-1)	LOWER STATE		UPPER STATE	
2752.000		V = 0	,J=	V=	,J=
	23.800				
2775.800		V = 0	,J=	V=	,J=
	23.200				
2799.000		V = 0	,J=	V=	,J=
	22.600				
2821.600		V = 0	,J=	V=	,J=
	22.000				
2843.600		V = 0	,J=	V=	,J=
	21.500				
2865.100		V = 0	,J=	V=	,J=

THE MISSING LINE, CENTERED BETWEEN THE TWO NEAREST PEAKS, IS AT 2885.650 CM-1

2906.200		V = 0	,J=	V=	,J=
	19.700				
2925.900		V = 0	,J=	V=	,J=
	20.000				
2945.900		V = 0	,J=	V=	,J=
	17.400				
2963.300		V = 0	,J=	V=	,J=
	17.700				
2981.000		V = 0	,J=	V=	,J=
	17.000				
2998.000		V = 0	,J=	V=	,J=

AVERAGE ENERGY DIFFERENCE DUE TO ROTATION = 20.50000 CM-1 = 4.071925E − 15 ERGS
ENERGY DIFFERENCE BASED ON TWO CENTRAL PEAKS = 20.55000 CM-1 = 4.081857E
− 15 ERGS
ALL QUANTITIES BELOW ARE BASED ON THE AVERAGE ENERGY
MOMENT OF INERTIA = 2.730806E − 40 G.CM**2
INTERNUCLEAR DISTANCE = 1.3000624E − 08 CM = 1.3006 ANGSTROMS
ROTATIONAL CHARACTERISTIC TEMPERATURE = 14.747 DEGREES KELVIN
ROTATIONAL PARTITION FUNCTION AT 298.150 DEGREES KELVIN = 2.0216869E + 01
ROTATIONAL ENERGY AT 298.150 DEGREES KELVIN = 4.1160800E − 14 ERGS/MOLECULE
= 5.9227625E − 01 KCAL/MOLE
ROTATIONAL ENTROPY AT 298.150 DEGREES KELVIN = 5.5311573E − 16 ERGS/MOLECULE.
DEGREE KELVIN
= 7.9589636E + 00 CAL/MOLE.DEGREE
KELVIN
ROTATIONAL SYMMETRY NUMBER = 1

FUNDAMENTAL VIBRATION FREQUENCY = 8.650962E + 13 CYCLES/SEC
ZERO POINT ENERGY = 2.8658907E − 13 ERGS/MOLECULE
= 4.1238241E + 00 KCAL/MOLE
VIBRATIONAL CHARACTERISTIC TEMPERATURE = 4151.840 DEGREES KELVIN
VIBRATIONAL FORCE CONSTANT = 5.513297E − 09 DYNES/CM
VIBRATIONAL PARTITION FUNCTION AT 298.150 DEGREES KELVIN = 9.465665E − 04
VIBRATIONAL ENERGY AT 298.150 DEGREES KELVIN = 2.8658958E − 13 ERGS/MOLECULE
= 4.1238314E + 00 KCAL/MOLE
VIBRATIONAL ENTROPY AT 298.150 DEGREES KELVIN = 1.8200000E − 21 ERGS/MOLECULE.
DEGREE KELVIN
= 2.6188576E − 05 CAL/MOLE.DEGREE
KELVIN

will be found with the values listed by the National Bureau of Standards[9] which are

$$H_{298} - H_0 = 2066 \text{ cal mole}^{-1}$$

$$S_{298} = 44.646 \text{ cal deg}^{-1} \text{ mole}^{-1}$$

$$C_{P298} = 6.96 \text{ cal deg}^{-1} \text{ mole}^{-1}$$

It may be interesting to see how the data of Figure 8-4 have been calculated and printed out by a digital computer. The computer print-out is shown in Table 8-1.

::

THE ENTROPY OF MIXING VERY SIMILAR MOLECULES OR IONS

In Chapter 4 we derived an equation, 4-16, in terms of the numbers of moles and the mole fractions for the entropy increase upon mixing of ideal gases. We can derive the same equation through 8-2. The name thermodynamic probability for W is not preferred by all authors; Pitzer and Brewer, for example, prefer *multiplicity*. However, for a given system the value of W for a particular distribution is proportional to the ordinary probability of finding that particular distribution. The normalizing factor would be the sum of the W's for all the possible distributions. Thus for the ordinary probability $\mathscr{P}_i$ of the ith distribution we can write

$$\mathscr{P}_i = \frac{W_i}{\sum W} \tag{8-71}$$

This would lead to

$$\frac{\mathscr{P}_i}{\mathscr{P}_j} = \frac{W_i}{W_j} \tag{8-72}$$

Figure 8-5 is similar to Figure 4-3 except that we have replaced the semipermeable pistons by an impermeable partition to keep the gases

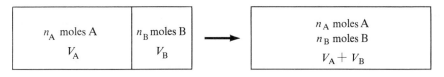

FIGURE 8-5

[9] D. D. Wagman, W. H. Evans, I. Halow, V. B. Parker, S. M. Bailey, and R. H. Schum, "Selected Values of Chemical Thermodynamic Properties," Part I, Technical Note 270-1 (1966).

A and B separate before mixing. From 8-2

$$\Delta S = S_2 - S_1 = k \ln \frac{W_2}{W_1} \tag{8-73}$$

or

$$\Delta S = k \ln \frac{\mathscr{P}_2}{\mathscr{P}_1} \tag{8-74}$$

Let us calculate $\mathscr{P}$ for the mixed state, $\mathscr{P}_2$, and for the unmixed state, $\mathscr{P}_1$.

The probability of finding each molecule of A and each molecule of B somewhere in the total volume, $V_A + V_B$, is unity. Hence $\mathscr{P}_2 = 1$. The probability of finding a particular molecule of A in V_A is $V_A \div (V_A + V_B)$. The probability of finding all the A molecules in V_A is $(V_A)^{n_A} \div (V_A + V_B)^{n_A}$. A similar expression applies to the probability of finding all B molecules in V_B. Hence

$$\frac{\mathscr{P}_2}{\mathscr{P}_1} = \frac{(V_A + V_B)^{(n_A+n_B)}}{(V_A)^{n_A}(V_B)^{n_B}} \tag{8-75}$$

and

$$\Delta S = n_A k \ln \frac{V_A + V_B}{V_A} + n_B k \ln \frac{V_A + V_B}{V_B} \tag{8-76}$$

If we now use the n's to mean numbers of moles instead of numbers of molecules, k becomes R and 8-76 becomes 4-14 which led simply to the result

$$\Delta S_{mix} = -R(n_A \ln X_A + n_B \ln X_B) \tag{4-16}$$

We can use similar reasoning to get an expression for the entropy of mixing of liquid and solid solutions of very similar molecules or ions. A nearly perfect example would be a solution of isotopes. For such condensed systems we can consider that there are $n_A + n_B$ sites that can be occupied by the molecules and that because of the similarity, there will be no preferred occupancy of a particular site by A or B. We can use Figure 8-5 to represent unmixed and mixed states except that the volumes will not enter into the present derivation. Let us calculate the W's for the unmixed and mixed states. For the unmixed state $W_1 = 1$ because permutations among indistinguishable molecules do not give rise to new configurations (all A's are indistinguishable from each other and all B's are indistinguishable from each other). For the randomly mixed state the number of configurations is the total number of permutations of the $n_A + n_B$ molecules divided by the number of permutations among indistinguishable molecules, namely,

$$W_2 = \frac{(n_A + n_B)!}{n_A! \, n_B!} \tag{8-77}$$

and

$$\Delta S_{mix} = k \ln \frac{W_2}{W_1} = k \ln W_2 = k \ln \frac{(n_A + n_B)!}{n_A! \, n_B!} \tag{8-78}$$

PROBLEM 8-30 By the use of Stirling's approximation, 8-11, show that 8-78 leads to 8-79 in which the *n*'s mean numbers of moles rather than numbers of molecules

$$\Delta S_{mix} = -R(n_A \ln X_A + n_B \ln X_B) \tag{8-79}$$

Note that 8-79 is identical in form to 4-16. By considering the derivation of 8-79 note that it can be generalized to

$$\Delta S_{mix} = -R(n_A \ln X_A + n_B \ln X_B + \cdots) \tag{8-80}$$

::

In the case of fused or solid solutions of a pair of salts such as silver chloride and silver bromide 8-79 applies because the two anions are sufficiently alike. In other cases, solid solutions do not exist and the liquid solutions, especially near their freezing points, do not have completely random distributions. In such liquids the order is like that in crystals but of shorter range.

PROBLEM 8-31 The high-temperature equilibrium mixture of ortho and para hydrogen has a 3 to 1 ratio of ortho to para. Calculate the entropy of mixing present in this equilibrium solution. This result should be valid for gaseous, liquid, or solid solutions.

::

THE ENTROPY OF VERY COLD CRYSTALS

In the preceding paragraph it was stated that the entropy of mixing of solid solutions of very similar substances can be treated by the same equations (8-79 or 8-80) that apply to liquid solutions. This fact should be true to as low temperatures as the assumptions in the derivation of 8-80 hold. Eastman and Milner (reference cited in Chapter 7) investigated the case of solid solutions of AgCl and AgBr and found 8-79 applicable to 15 °K and found no reason to believe that the system would show any departure from this relationship at any lower temperatures. The critical assumption in the derivation of this relationship is the lack of any energy change corresponding to a substitution of A for B or vice versa at any of the molecular, anionic or cationic sites in the crystal. Philosophically one can doubt that this assumption would be exactly true for any case in which A and B are distinguishable. Practically, then, it is a question of the strength of the tendency of the system to assume the theoretically preferred structure compared to the difficulty of its doing so.

As the temperature falls, the tendency to assume the preferred structure presumably would remain more or less constant. The difficulty of its doing so involves two opposing factors. One of these, the decreasing thermal agitation, would be conductive to the maintenance of the preferred structure, once attained. But this decrease in agitation—a decrease in the mobility of ions or molecules—impedes the attainment of the preferred structure. If one could bring the temperature of any system to absolute zero and could wait long enough, a unique ordering of the molecules or ions might be expected. The fact is, however, that in some cases the required closeness of approach to $T = 0$ or the length of time required at very low temperatures for the system to attain, or nearly attain, the unique preferred structure, or both, cannot be provided experimentally. But when mixtures of very similar substances are excluded, cases of crystals failing to achieve a unique order as $T \rightarrow 0$ are a very small minority.

Let us first consider the common case, i.e., that of a crystal which at very low temperatures has a unique structure and does not have to be considered as any kind of a randomly mixed solid solution. As $T \rightarrow 0$, the thermodynamic probability will be given by the same expression as for the ideal gas, namely,

$$W_0 = g_0{}^n \qquad (8\text{-}38)$$

and the entropy per mole, accordingly, by

$$S_0 = R \ln g_0 \qquad (8\text{-}39)$$

In most cases $g_0 = 1$ and hence

$$S_0 = 0 \qquad (8\text{-}81)$$

This is the basis of the practical Third Law.

An instructive pair of cases of conformance to the practical Third Law are those of rhombic and monoclinic sulfur. By comparison of the entropy of the rhombic-monoclinic transition at the equilibrium temperature (368.5 °K) calculated by the Second-Law equation

$$\Delta S = \frac{\Delta H}{T} \qquad \begin{bmatrix} \text{For a phase} \\ \text{transition at} \\ \text{the equilibrium} \\ \text{temperature} \end{bmatrix} \qquad (4\text{-}9)$$

with the entropy increments measured for heating the two forms from 0 °K to 368.5 °K, it is found within experimental error that both forms have the same entropy (to which we would assign a value of zero) at 0 °K.

The possibly illuminating fact about this example is that it is not necessary for the sulfur to be in its most stable form at 0 °K for it to have zero entropy, only that it have a unique, perfectly ordered structure. On second thought, however, it should be evident that that is exactly what we found in the Ag, Cl_2, AgCl case, for there we found that to both the unstable system Ag + Cl_2 and to the stable system AgCl a value of $S_0 = 0$ could be assigned in consistency with all the experimental facts.

The case of the ice crystal is typical of several known exceptions to the practical Third Law which include also the cases of CO, N_2O, and NO. These compounds are said to have a residual entropy at the lowest attainable temperatures. This means that for these substances the entropy at some temperature such as 25 °C calculated from heat capacity measurements extrapolated to 0 °K, and assuming $S_0 = 0$, is less than the entropy at the same temperature calculated statistically from spectroscopic data or as determined by the application of Second-Law relationships to a chemical reaction in which the entropy of the other reactants and products is not in question. The inconsistency is formally removed by saying that for these substances $S_0 \neq 0$. The question is, how can a nonzero assignment to S_0 in these cases be brought into accord with the statistical basis of entropy which we have discussed.

      The answer is closely related to the preceding comments about solid solutions at very low temperature. There remains in the H_2O, CO, N_2O, NO and a relatively few other crystals a measure of randomness of structure at the lowest attainable temperatures and for the longest feasible periods of observation. By randomness is meant the fact that there are numbers of essentially equally probable structures, that is, that W is significantly larger than unity.

      In 1935 Linus Pauling[10] proposed a model of the ice crystal at very low temperatures which would account very largely for the observed discrepancy in the case of water and of heavy water between the "Third-Law" entropy and the entropy obtained statistically from spectra or obtained by Second-Law treatments of equilibrium or electrochemical data. The Third-Law entropies (at, say 298 °K) are 0.82 and 0.77 cal deg^{-1} mole^{-1} less for H_2O and D_2O, respectively, than the entropies calculated in the other ways. These facts can be stated by saying that H_2O and D_2O have residual entropies at 0 °K of 0.82 and 0.77 cal deg^{-1} mole^{-1} respectively.

      To a great extent we abbreviate the discussion of the crystal structure of ice. It is perhaps sufficient to think of it as a lattice of oxygen atoms, each oxygen atom being surrounded roughly tetrahedrally by four hydrogen atoms and each hydrogen atom lying between but not equidistant from a pair of oxygen atoms. Figure 8-6 shows these two features of the ice

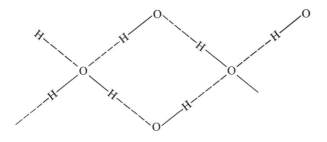

FIGURE 8-6

[10] See Linus Pauling, *The Nature of The Chemical Bond*, Third Edition, Cornell University Press (1960).

crystal. Each pair of O atoms is connected through an H atom by a pair of bonds, one a stronger shorter (covalent) bond and one a longer weaker (hydrogen) bond. Each O atom can have only two strong short bonds and only two longer weaker ones. It is assumed that the interaction among adjacent water molecules is not sufficient to stabilize any configuration with respect to the placement of the long and short bonds but that as the temperature is lowered to near 0 °K, complete randomness in this respect is "frozen in."

Let us calculate W, the number of equally probable configurations for this model. Each H has two possible locations with respect to the pair of O atoms between which it lies. There are $2L$ H atoms per mole (L is Avagadro's number). But not all arrangements conceivable on this basis are chemically possible because of the restriction that each O atom must have 2 long and 2 short bonds to H atoms. If we tabulate the number of arrangements around each O atom conceivable without regard to the chemical restriction we get:

Configuration	Number of Arrangements
4 long, 0 short	1
3 long, 1 short	4
2 long, 2 short	6
1 long, 3 short	4
0 long, 4 short	1
Total	16

PROBLEM 8-32 Show that the total number of conceivable arrangements without regard to the chemical restriction is 2^{2L} per mole of ice. By reference to the above tabulation show that only $\frac{3}{8}$ of the total conceivable arrangements around each O are chemically possible. Show that, then, for a mole of ice $W = (\frac{3}{8})^L 2^{2L}$ and that this fact leads to a value for S for ice at 0 °K which is near the experimental values for H_2O and for D_2O.

::

PROBLEM 8-33 The model that gives the observed residual entropy at 0 °K for N_2O and CO is one in which the molecules in the crystal have two equally likely orientations, i.e., NNO or ONN and CO or OC. Show that this model leads to a residual entropy for these compounds of $R \ln 2$ per mole, a value in agreement with experimental results.

::

The last case involving residual entropy which we consider is that of hydrogen. The proper value to use for the entropy of hydrogen was for some time a matter of uncertainty and disagreement. A detailed consideration of this problem was presented by W. F. Giauque in 1930.[11] The problem depends largely on the fact that nuclear spin makes it possible for two kinds of H_2 molecules to exist. Ortho hydrogen is H_2 with the nuclear spins parallel,

[11] W. F. Giauque, *J. Am. Chem. Soc.*, **52**, 4823 (1930).

para hydrogen has the spins antiparallel. At ordinary temperatures the two forms are more or less readily interconvertible; the equilibrium mixture at room temperature and above is $\frac{3}{4}$ ortho and $\frac{1}{4}$ para. At low temperatures, however, where the equilibrium shifts toward para H_2 (as $T \to 0$, if equilibrium between the forms were maintained, the H_2 would become entirely para H_2) the rate of interconversion in the absence of a catalyst becomes extremely small. Heat capacity measurements, as usually carried out, even extrapolated to 0 °K would refer to the 3 to 1 ortho-para mixture.

In the case of HCl we dealt with an unsymmetrical molecule for which the quantum rules allow all integral values for the rotational quantum number J. In the case of a symmetrical molecule such as H_2 one may use only odd or only even values of J. It is even values in the case of para H_2 and odd values for ortho H_2. Thus, the lowest rotational state permitted for ortho H_2 would be $J = 1$ and from 8-49 we would write $g_0 = 3$ for ortho H_2. Another factor which must be taken into account is the entropy of mixing of ortho and para H_2. The value of W associated with nuclear spin orientation can usually be left out of consideration in chemical problems for the same reason that the entropy of isotope mixing can, but it is usually discussed in connection with the hydrogen problem. We do not resolve the problem except to say that for chemical purposes the standard entropy of H_2 at 25 °C is 31.208 cal deg⁻¹ mole⁻¹ and that good reviews of the problem have been given by a number of authors such as Dole[12] and Pitzer and Brewer.[13]

ADDITIONAL PROBLEMS

PROBLEM 8-34 Identify the transitions corresponding to the various peaks in the infrared spectrum of HCl by filling in the numbers for the V's and J's in Table 8-1.

PROBLEM 8-35 Using the data in Figure 8-4 (and the molecular weight of HCl) calculate

$$\frac{G° - H_0°}{T}$$

for HCl at 1000 °K. Assume that excited electronic states can be ignored. Pitzer and Brewer give a value of 46.16 cal deg⁻¹ mole⁻¹.

[12] Malcolm Dole, *Introduction to Statistical Thermodynamics*, Prentice-Hall, 1954.

[13] Lewis and Randall, *Thermodynamics*, revised by K. S. Pitzer and L. Brewer, McGraw-Hill, 1961.

9

Chemical Equilibrium and the Chemical Potential in Ideal Gases

A majority of the discussion in the preceding chapter was related to statistical methods of evaluating thermodynamic properties of pure substances in the ideal gas state. The evaluation of these properties for pure condensed phases can also be done, but it is not practical in many cases to do it by purely statistical methods. It can be done, for example, through the practical Third Law. It should now be evident that this type of treatment combines an essentially statistical basis for the zero entropy assumed at $0\,°K$ and an essentially Second Law (nonstatistical) calculation of the entropy increment from $0\,°K$ to the temperature of interest. Or it can be done in many cases by a statistical treatment of spectroscopic data for the substance in the ideal gas state together with the use of Second-Law methods to calculate the increments in the functions corresponding to compression and condensation of the gas to the liquid or solid state. Moreover, we should not lose sight of the fact that thermodynamic properties of chemical systems whether gas, liquid, or solid have been and can be measured and used without resort to statistical methods at all.

We are about to develop the thermodynamic basis of the quantitative relationships which prevail in chemical equilibria. It is convenient to start with the idea that substances in specified states can be assigned consistent sets of values for their thermodynamic properties. By consistent sets of values we mean that although there is an arbitrary base line for E, H, A, G, and even S in some cases (as discussed briefly in connection with the entropy of H_2, for example), differences among these values will be equal to the differences that can be measured in chemical reactions.

The following equation can be considered the generalized

expression for a chemical reaction:

$$bB + cC + \cdots \rightarrow hH + iI + \cdots \qquad (9\text{-}1)$$

In 9-1 the lower case letters are the stoichiometric numbers and the capitals represent the formulas for the elements or compounds involved in the reaction.

OPEN SYSTEMS

Consider a mixture of B, C, H, I, and so forth, maintained at constant pressure and temperature. Reaction among these substances is equivalent, as far as the properties of the system are concerned, to putting into the mixture certain amounts of the products and taking out of it certain amounts of the reactants. A degree-of-advancement parameter ξ[1] can be implicitly defined by noting that for an infinitesimal amount of reaction the molar increments of the various substances will be

$$dn_B = -b\,d\xi, \qquad dn_C = -c\,d\xi, \qquad dn_H = h\,d\xi, \qquad dn_I = i\,d\xi \quad (9\text{-}2)$$

Perhaps an example will be useful. For the reaction $N_2 + 3H_2 \rightarrow 2NH_3 - d\xi$ is the number of moles of N_2 formed, ($d\xi$ is the number of moles of N_2 used up) when the reaction proceeds by an infinitesimal amount. Likewise for the infinitesimal advancement of the reaction $dn_{H_2} = -3\,d\xi$ and $dn_{NH_3} = 2\,d\xi$. If the degree of advancement is ξ (finite),

$$\Delta n_{N_2} = -\xi, \qquad \Delta n_{H_2} = -3\xi, \qquad \Delta n_{NH_3} = 2\xi$$

We have as the criterion for equilibrium in the constant T and P system

$$\left(\frac{\partial G}{\partial \xi}\right)_{T,P} = 0 \qquad (5\text{-}39)$$

and the corresponding criterion for the possibility of spontaneous change in such a system is

$$\left(\frac{\partial G}{\partial \xi}\right)_{T,P} < 0 \qquad (9\text{-}3)$$

Let us denote by μ_B the quantity

$$\mu_B = \left(\frac{\partial G}{\partial n_B}\right)_{T,P,n_C,n_D\cdots n_H,n_I\cdots} \qquad (9\text{-}4)$$

Then by considering the chemical reaction to be equivalent to successively adding small stoichiometric amounts of H, I, etc., and successively taking away small stoichiometric amounts of B, C, etc. from our reaction mixture at

[1] Theophile de Donder developed the use of the *degree of advancement* parameter. See T. de Donder and P. Van Rysselberghe, *The Thermodynamic Theory of Affinity*, Stanford University Press, 1936.

constant T and P we get

$$dG = (-b\mu_B - c\mu_C - \cdots + h\mu_H + i\mu_I + \cdots) \, d\xi \qquad (9\text{-}5)$$

By combining 9-2, 9-3, and 9-5 we get as criteria for equilibrium and for the possibility of reaction

$$\widetilde{\Delta\mu} \equiv \frac{\partial G}{\partial \xi} = -b\mu_B - c\mu_C - \cdots + h\mu_H + i\mu_I + \cdots \leqq 0 \qquad (9\text{-}6)$$

where the equality is the necessary condition for equilibrium and the inequality the condition for the possibility of spontaneous change.

THE μ's AND CHEMICAL EQUILIBRIUM

To make 9-6 useful we must develop ways of evaluating the μ's in terms of directly measurable properties. It should be noted that the derivation of 9-6 has not involved the specification of any particular kinds of systems—ideal gases or homogeneous mixtures, for example. It is, therefore, a generally valid thermodynamic equation and applicable to any chemical reaction system at constant T and P. The easiest case and the one we now consider is that of a reaction among ideal gases.

The Gibbs energy, $\bar{G}_B$,[2] per mole of B in a nonstandard state can be written in terms of its value $\bar{G}_B^\circ$ in a standard state, together with a term equal to the change in $\bar{G}_B$ corresponding to the change from standard to nonstandard state. The definition of G is

$$G \equiv H - TS \qquad (5\text{-}8)$$

from which we get

$$dG = V \, dP - S \, dT \qquad (5\text{-}21)$$

The standard state of an ideal gas is the gas at 1 atm and the temperature of interest. Thus, at a particular temperature

$$\bar{G}_B = \bar{G}_B^\circ + \int_{P_B=1}^{P_B} V \, dP_B \qquad (9\text{-}7)$$

which for the ideal gas gives

$$\bar{G}_B = \bar{G}_B^\circ + RT \ln \frac{P_B}{P_B^\circ} \qquad (9\text{-}8)$$

But we have stated that conventionally the standard state of an ideal gas is the gas at 1 atm, i.e., $P_B^\circ = 1$ atm. Hence, if P_B is expressed in atmospheres, and remembering that μ_B and $\bar{G}_B$ are two different symbols for the same quantity, we can write

$$\mu_B = \bar{G}_B^\circ + RT \ln P_B \qquad (9\text{-}9)$$

or

$$\mu_B = \mu_B^\circ + RT \ln P_B \qquad (9\text{-}10)$$

[2] The symbol $\bar{G}_B$ and analogous symbols such as $\bar{H}_B$ and $\bar{V}_B$ in general are used to denote partial molal properties, which are discussed in some detail later. For ideal gases, however, the molal and the partial molal values of these properties are the same.

PROBLEM 9-1 Show that by putting 9-10 into the equality of 9-6 one can get the following relation for the equilibrium partial pressures:

$$b\bar{G}_B^\circ + c\bar{G}_C^\circ + \cdots - h\bar{G}_H^\circ - i\bar{G}_I^\circ - \cdots = RT \ln \frac{P_H{}^h P_I{}^i \cdots}{P_B{}^b P_C{}^c \cdots} \quad (9\text{-}11)$$

which is equivalent to

$$\widetilde{\Delta G}^\circ = -RT \ln K_P \quad (9\text{-}12)$$

where, if P_B, P_C, P_H, P_I are the partial pressures in an equilibrium mixture

$$K_P \equiv \frac{P_H{}^h P_I{}^i \cdots}{P_B{}^b P_C{}^c \cdots} \quad (9\text{-}13)$$

::

The result of Problem 9-1 is the basic law of chemical equilibrium among ideal gases. What we have done in that problem is first to derive 9-11 from our Gibbs energy criterion of equilibrium 9-6. Equation 9-11 shows a relation between $\widetilde{\Delta G}^\circ$ (which we observe the left-hand side of 9-11 to be) and a certain function of the equilibrium pressures to which function we give the name (symbol) K_P (9-13). But inasmuch as ΔG° for a given reaction at a given temperature is a constant, then the function of the equilibrium pressures which we designate as K_P must be a constant for a given reaction at a given temperature. One point should perhaps be emphasized. It is that because the standard state of an ideal gas is 1 atm, the lower limit of the integral in 9-7 was taken as 1 (not 760, for example). This means that the values of the P's in 9-11 and 9-13 should be expressed in atm.

PROBLEM 9-2 The pressures which appear in 9-11 and 9-13 are the partial pressures in the equilibrium mixture. Let us denote by P_B' etc., arbitrarily chosen pressures. An overall change in state as well as a possible way of bringing it about are indicated in Figure 9-1. Show that 9-14 follows from other equations previously derived.

$$\Delta G = \Delta G^\circ + RT \ln \frac{P_H'^h P_I'^i}{P_B'^b P_C'^c} \quad (9\text{-}14)$$

::

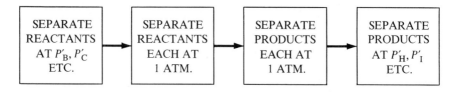

FIGURE 9-1

PROBLEM 9-3 In Problem 9-2 we were considering a reaction starting with pure substances at arbitrary pressures, these substances being then somehow converted completely into pure reaction products also at arbitrary pressures. In theory this could be done with cylinders, pistons, and appropriate semipermeable membranes. An easier experiment to do and one possibly a little more difficult to analyze is to start with reactants not at chemical equilibrium and then to let the system undergo chemical reaction to some extent. Figure 9-2 specifies the problem. If we denote the value of μ_B in the first state by μ_B and in the second state by μ_B' and also write $\Delta\mu_B = \mu_B' - \mu_B$ show (see 9-6) that

$$\Delta G = n_B \, \Delta\mu_B + n_C \, \Delta\mu_C + \cdots + n_H \, \Delta\mu_H + n_I \, \Delta\mu_I + \cdots + \xi \, \Delta\tilde{\mu}' \tag{9-15}$$

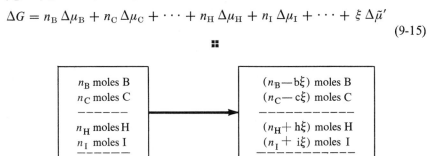

FIGURE 9-2

PROBLEM 9-4 Equation 9-15, like 9-6, was derived without specifying the nature of the reacting substances and can therefore be used in any case in which it is possible to evaluate the μ's. It should also be borne in mind that the derivation did not specify that the second state was one at chemical equilibrium. Show that if the system shown in Figure 9-2 is one made up of ideal gases and if in the second state the system *is* at chemical equilibrium:

$$\Delta G = RT\left(n_B \ln \frac{P_B'}{P_B} + n_C \ln \frac{P_C'}{P_C} + n_H \ln \frac{P_H'}{P_H} + \cdots\right) \tag{9-16}$$

There is no ξ in 9-16. Why?

PROBLEM 9-5 Show that as the pressure of an ideal gas approaches zero, $\mu \to -\infty$. We may inquire, therefore, what happens to 9-15 if there is no H, I, etc., in the reaction mixture at the start. What is the value, under these conditions, of terms such as $n_H\mu_H$ in which the first factor is 0 and the second $-\infty$? Writing $\mu_H = \mu_H^\circ + RT \ln P_H$ and also $P_H = (n_H/n)P$ where n is the total moles in the mixture and P is the total pressure, show that as $n_H \to 0$, $n_H\mu_H \to 0$. One way that this can be done is to expand $\ln(n_H/n)$ into a series.

Theoretically it would be possible to eliminate all the equilibrium pressures which appear in 9-16 and to replace them by expressions in terms of the stoichiometric numbers, the equilibrium constant and the total pressure in the equilibrium mixture. Such an expression for the general case gets very cumbersome, however, and 9-16 is probably about the simplest starting point for the calculation of ΔG for a reaction among ideal gases which goes to equilibrium.

PROBLEM 9-6   A very simple example of the calculation of ΔG for a chemical reaction allowed to go to equilibrium would be for an ideal gas reaction of the type $B + C \rightarrow H + I$. Suppose we start with 1 mole each of B and C, separated and in their standard states, and allow them to react isothermally to an equilibrium under a total pressure of 1 atm. Show that for this change in state

$$\Delta G = 2RT \ln \frac{(\sqrt{K_P} - 1)}{2(K_P - 1)}$$

By the use of l'Hôpital's rule or otherwise show that if $K_P = 1$, $\Delta G = -4RT \ln 2$. For the case that $K_P \gg 1$ show that $\Delta G = \Delta G° - 2RT \ln 2$. How do you account for the $2RT \ln 2$?

::

One further equation can now easily be added to our set concerning chemical equilibria among ideal gases. We have previously mentioned the Gibbs–Hemholtz equation in several connections. The form of it which is most convenient for our present needs is

$$\frac{\partial \left(\dfrac{\Delta G}{T}\right)}{\partial T} = \frac{-\Delta H}{T^2} \tag{5-35}$$

If we apply 5-35 to 9-12 we get

$$d \ln K_P = \frac{\Delta H°}{RT^2} dT \tag{9-17}$$

For the case in which it is a satisfactory approximation to treat $\Delta H°$ as independent of T, the integrated expression becomes

$$\ln \frac{(K_P)_2}{(K_P)_1} = \frac{\Delta H°}{R} \left(\frac{T_2 - T_1}{T_1 T_2}\right) \tag{9-18}$$

If the temperature range is too long to make constancy of $\Delta H°$ a satisfactory approximation, one can use more elaborate methods such as the one used to derive 5-37.

Later we extend our study of chemical equilibria to systems other than ideal gases. There are two related reasons that ideal gases have been especially easy to treat. One is that because of the simple equation of state, it is easy to calculate the μ's at any pressure. The second reason is that ideal gases in mixtures have the same values of E, H, S, A, and G as they would have separately at the same temperature and at pressures equal to their partial pressures in the mixture. There is no ΔH of mixing and the entropy of mixing (as we showed in deriving 4-4) is just that corresponding to changing the pressure of each component gas to its partial pressure in the mixture. Therefore, there is no change in the molal E, H, S, A, or G of an ideal gas when, at a pressure equal to its partial pressure in the mixture, it is introduced into the mixture.

THE CHEMICAL POTENTIAL

Although we may find that devious ways must be followed in dealing with some kinds of systems to evaluate μ for the various components, μ is adequately defined for all cases by equations of the form of 9-4. Partial derivatives of the form $(\partial X/\partial n_{\mathrm{B}})_{T,P,n_{\mathrm{C}},n_{\mathrm{D}}}\ldots$ where X is any extensive property of a system are called *partial molal properties*. As defined by 9-4, μ_{B} is the partial molal Gibbs energy of component B in a mixture of B, C, etc., and for an ideal gas has, as we have stated, the same value as the molal Gibbs energy of pure B at a pressure corresponding to its partial pressure in the mixture.

μ_{B} is called the chemical potential of B, and $\Delta\tilde{\mu}$ is called the reaction potential. The logic of the name *chemical potential* appears more completely when we consider phase equilibria, but 9-6 suggests that it is a reasonable name. The question might be asked why call it anything other than the partial molal free energy (or partial molal Gibbs energy). The answer, such as it is, is that μ can be defined equally correctly but usually not so usefully in terms not involving G. The total increment in any thermodynamic property of a system can be expressed as the sum of all the increments corresponding to changes in all of the independent variables affecting the system. For a closed system with only the usual T and PV interactions with the surroundings there are two independent variables which two may be any pair such as PT, PV, TP, ES, and so on. With open systems, which we have been considering in this chapter, to the two independent variables listed in the preceding sentence must be added the variation in number of moles for each component of the system. For a closed system, for example,

$$dA = -P\,dV - S\,dT \tag{5-20}$$

and

$$dG = V\,dP - S\,dT \tag{5-21}$$

For open systems we would write

$$dA = -P \, dV - S \, dT + \left(\frac{\partial A}{\partial n_1}\right)_{T,V,n_2\cdots} dn_1 + \left(\frac{\partial A}{\partial n_2}\right)_{T,V,n_1,n_3,\dots} dn_2 + \cdots$$

$$(9\text{-}19)$$

and

$$dG = V \, dP - S \, dT + \left(\frac{\partial G}{\partial n_1}\right)_{T,P,n_2\cdots} dn_1 + \left(\frac{\partial G}{\partial n_2}\right)_{T,P,n_1,n_3,\dots} dn_2 + \cdots$$

$$(9\text{-}20)$$

in which n_1, n_2, etc., refer to the numbers of moles of components 1, 2, and so on.

From the defining equation 9-4, 9-20 can be rewritten

$$dG = V \, dP - S \, dT + \mu_1 \, dn_1 + \mu_2 \, dn_2 + \cdots \qquad (9\text{-}21)$$

PROBLEM 9-7 By comparing the defining equations for A and G (5-7 and 5-8) it is easy to show that $G = A + PV$ and hence that $dG = dA + P \, dV + V \, dP$. By substituting this last expression for dG into 9-23 show that

$$\mu_1 = \left(\frac{\partial A}{\partial n_1}\right)_{T,V,n_2\cdots} \qquad (9\text{-}22)$$

Thus, 9-22 is as correct a definition of μ_1 as is 9-4.

::

PROBLEM 9-8 By treatments similar to that of Problem 9-7 show that other correct expressions for μ_1 are the following

$$\mu_1 = \left(\frac{\partial E}{\partial n_1}\right)_{S,V,n_2,n_3\cdots} \qquad (9\text{-}23)$$

$$\mu_1 = \left(\frac{\partial H}{\partial n_1}\right)_{S,P,n_2,n_3\cdots} \qquad (9\text{-}24)$$

$$\mu_1 = -T\left(\frac{\partial S}{\partial n_1}\right)_{E,V,n_2,n_3\cdots} \qquad (9\text{-}25)$$

$$\mu_1 = -T\left(\frac{\partial S}{\partial n_1}\right)_{H,P,n_2,n_3\cdots} \qquad (9\text{-}26)$$

Note that the μ_1 in all of these equations is the *same* μ_1. It is not a different-but-analogous kind of chemical potential.

::

ADDITIONAL PROBLEMS

PROBLEM 9-9   For the reaction

$$SO_2(g) + \tfrac{1}{2}O_2(g) \rightarrow SO_3(g)$$

$\Delta G° = -16.73$ kcal at 25 °C. Calculate the partial pressure of each gas in an equilibrium mixture of total pressure of 0.1 atm and at 25 °C if the equilibrium mixture is obtained by (*a*) catalyzing the decomposition of pure SO_3, (*b*) catalyzing the reaction between an equimolar mixture of SO_2 and O_2.

::

PROBLEM 9-10   Assume that the decomposition of HNO_3 can be represented by the following equation

$$4HNO_3(g) \rightarrow 4NO_2(g) + 2H_2O(g) + O_2(g)$$

Assume also that at the temperature and pressure of interest the reaction approaches equilibrium and that any side reactions can be ignored. (*a*) Show that if one starts with pure HNO_3

$$K_P - \frac{1024P_{O_2}^7}{(P - 7P_{O_2})^4} = 0$$

where K_P is the equilibrium constant, P is the total pressure and P_{O_2} is the pressure of oxygen in the equilibrium mixture. (*b*) Assume that at the temperature of interest $K_P = 1$. Solve for the value of P_{O_2} at this temperature and under a total pressure of 1 atm. Hint: by considering stoichiometric relations put limits on the values that P_{O_2} could possibly have, then guess values until one is found which very closely satisfies the equation. Only a few guesses should be needed to evaluate P_{O_2} to 3 places, especially if one bases each guess on the results of the previous ones.

::

PROBLEM 9-11   Suppose that at 25° and 1 atm an equimolar mixture of SO_2 and O_2 are allowed to react until the mixture is equimolar in SO_2 and SO_3. (*a*) What is ξ? (*b*) What is ΔG for this change in state? (See Problem 9-9.)

::

PROBLEM 9-12   For the reaction

$$2SO_2(g) + O_2(g) \rightarrow 2SO_3(g)$$

1 mole of SO_3 gas at 25° and 0.1 atm is heated in a constant-volume container to 200 °C. What will be the pressure in the container when the decomposition has gone to equilibrium? (See Problem 9-9.) What approximations were made in solving this problem?

10

Chemical Potentiometry

The correspondences between thermal potentiometry (thermometry) and chemical potentiometry are instructive and striking. We make several comparisons. First, the ideal gas was found to be the most convenient substance with respect to which to introduce quantitative considerations concerning temperature. The same thing is true of the chemical potential. Although in both cases much more general thermodynamic formulations of the potentials exist, the ideal gas remains in many cases the most convenient meter for accurate primary determinations of either the temperature or the chemical potentials of a system.

Second, either the temperature or the chemical potentials in a system can be measured by equilibrating the system with a primary or calibrated standard, the thermometer, or what we can call a chemical potentiometer. In neither case do we mix the meter with the system to be measured but we put them into a suitable kind of communication with each other. For measuring temperature, heat is able to flow between the system and the thermometer through a heat-conducting wall. For measuring chemical potentials, matter must be able to pass through phase boundaries or semipermeable walls.

Third, there are ways of measuring the temperature of some systems without having to equilibrate them with another system. One might think of these methods as ways of using the system as its own thermometer. The determination of the temperature of an ideal gas by measuring its pressure and volume is the simplest example. Another example is the determination of the absolute temperature of a reversible electrochemical cell without a thermometer as was done in Chapter 2.[1] Likewise we can determine

---

[1] In the course of the derivation we did use an arbitrary temperature scale, but that scale could have been defined as equal to the potential of the cell, for example.

chemical potentials of gases and of substances present in suitable electro-chemical systems without equilibration with an external chemical potentiometer.

Fourth, both temperature scales and scales of chemical potential are arbitrary. Scientists most commonly state temperatures in °C and most commonly use °K in calculations. Moreover, many thermodynamic formulas come out with terms in log T or e^T and either of these functions of T might have been given names and used as alternative temperature scales. So it is with chemical potentials. Two systems will be in chemical equilibrium with each other if for each component the partial molal free energies are equal, the vapor pressures are equal, the fugacities are equal, or if the activities (based on the same standard state) are equal. Thus, although of these only the partial molal free energy is given the name chemical potential, they all have mono-tonic relationships with each other and could be used as scales of chemical reactivity. The vapor pressure is the most tangible, but the other three measures of reactivity give simpler formulas.

THE ESCAPING TENDENCY

To see how a chemical potentiometer might be used let us first consider a two phase system in a constant-volume container composed of a liquid solution and a vapor, there being only one appreciably volatile component, B, in the solution and only B in the gas phase. The tendency of B to leave the solution is sometimes called its escaping tendency in the solution and its tendency to leave the vapor, its escaping tendency in the vapor. When the escaping tendency in the two phases is equal, there will be no net transfer and the system will be in equilibrium. As suggested in the preceding paragraph, the escaping tendency could be assigned a value in the vapor equal to its pressure and in the liquid equal to the equilibrium vapor pressure of B over the solution. But we can also assign consistent values to the escaping tendency in terms more directly related to the chemical potentials $\mu_B{}^l$ and $\mu_B{}^g$ of B in the liquid and in the vapor.

PROBLEM 10-1   Show that at equilibrium an equation entirely analogous to 5-39 must apply to the 2-phase system, namely

$$\left(\frac{\partial A}{\partial \xi}\right)_{T,V} = 0 \tag{10-1}$$

and that, in the light of 9-24

$$\mu_B{}^l = \mu_B{}^g \tag{10-2}$$

■■

Now, if we can evaluate $\mu_B{}^g$ at equilibrium we have a value of $\mu_B{}^l$ in the equilibrium solution. The simple case is that the vapor is sufficiently dilute so that it is justifiable to use 9-10

$$\mu_B = \mu_B^\circ + RT \ln P_B \qquad (9\text{-}10)$$

In that case the difference in the chemical potentials of B in solutions of two different concentrations would be related to the differences in vapor pressures by

$$\mu_B' - \mu_B'' = RT \ln \frac{P_B'}{P_B''} \qquad (10\text{-}3)$$

10-3 also, of course, relates the μ's of gaseous B at two different pressures whether or not the gas is in equilibrium with a liquid phase.

PROBLEM 10-2   The vapor pressure of water over a sulfuric acid solution containing 425 grams of H_2SO_4 per liter of solution is 12.2 torr at 20 °C. At this temperature the vapor pressure of pure water is 17.4 torr. Calculate in calories per mole the difference between μ for pure water and μ for the water in the solution at this temperature.

::

The preceding example has illustrated the importance of chemical potentials in gases not only for their own sakes but because gases can be used as chemical potentiometers for other systems. This usefulness is not restricted to ideal gases but applies to nonideal gases when their μ's can be determined. Such determinations are straight forward for pure nonideal gases if the P–V relationships are known at the temperature of interest. 9-7 is equivalent to

$$\mu = \mu^\circ + \int V\, dP \qquad (10\text{-}4)$$

where the integration is taken from the pressure and volume corresponding to the standard state of the gas to the P and V of the state for which μ is the chemical potential. We must now consider the question of standard states in more detail than we have done previously.

STANDARD STATES

The standard states which we have previously defined have been defined in terms of real states. Thus, the standard state of an ideal gas is its state under 1 atm pressure. Likewise, for purposes such as the tabulations of ΔH's or ΔG's of formation and reaction of pure liquid and solid substances, their standard states are usually defined as their real states under 1 atm pressure.

A different formalism is convenient and is usually adopted for the standard states of nonideal gases and gas mixtures and of liquid and solid solutions. In this formalism the emphasis is not so much on the standard state (which may be an imaginary one, the "hypothetical 1 molal solution," for example) as it is on the reference state which is usually the infinitely dilute solution or gas. This reference state is convenient because of simplified relationships which exist in very dilute gases (which approach ideal gas behavior) and in very dilute solutions.

We start the implicit definition of the standard state of a nonideal gas at the temperature T by defining a ratio f'/f'' by

$$\mu' - \mu'' \equiv RT \ln \frac{f'}{f''} \tag{10-5}$$

where the single and double primes refer to the same gas in two different states. A comparison of 10-3 and 10-5 indicates that the f's could be thought of as the pressures of the gas in the two states corrected in a way that will take account of the nonideality of the gas. We can further this interpretation by requiring that

$$\frac{f}{P} \to 1 \quad \text{as} \quad P \to 0 \tag{10-6}$$

f is called the *fugacity*, a name based on the fact that it, like the pressure itself or the chemical potential, is a measure of the escaping tendency of the gas. It is, as we later see, just a special representation of the more generally defined function, the activity.

PROBLEM 10-3   From 10-4 and 10-5 we get

$$RT \ln \frac{f'}{f''} = \int_{P''}^{P'} V \, dP \tag{10-7}$$

Then, thinking of the f's as corrected pressures, we subtract from 10-7 an equation like it but applicable to an ideal gas (which, however, is also just a mathematical identity at constant T), namely

$$RT \ln \frac{P'}{P''} = \int_{P''}^{P'} \frac{RT}{P} \, dP \quad \begin{bmatrix} \text{Ideal} \\ \text{gas} \end{bmatrix} \tag{10-8}$$

This gives us an expression in which the integral is the value of the correction for nonideal behavior. Show that the resulting relation can be written

$$RT \ln f' = RT \ln P' + RT \ln \frac{f''}{P''} + \int_{P''}^{P'} \left(V - \frac{RT}{P} \right) dP \tag{10-9}$$

Show, by letting P approach zero, that we can write for the value of f at any pressure P

$$\ln f = \ln P + \int_0^P \left(\frac{V}{RT} - \frac{1}{P} \right) dP \tag{10-10}$$

PROBLEM 10-4   Use 10-10 and 7-13 to calculate the pressure of $Cl_2(g)$ under which its fugacity is 1 atm at its boiling point, 239 °K. *Answer*: 1.03 atm.[2]

⁛

PROBLEM 10-5   By reviewing the derivation of 9-12 and the definition of *f*, show that if B, C, H, I, etc. of the generalized chemical reaction 9-1 are nonideal gases, the function of the fugacities which for a given temperature and reaction must be constant in all equilibrium mixtures is

$$\frac{f_H{}^h f_I{}^i \cdots}{f_B{}^b f_C{}^c \cdots} \quad \text{(symbol } K_f) \tag{10-11}$$

⁛

RAOULT'S AND HENRY'S LAWS

We can draw some conclusions about chemical equilibrium relations in liquid solutions by considering the possibility that a liquid reaction system and a vapor reaction system are in equilibrium with each other. It is helpful, perhaps, to think of a chemical reaction in which all the reactants and products are appreciably volatile so that all the components will be present in measurable amounts in both phases. However, the theory of our derivation does not require any minimum amount of any component in either phase, and the result will be valid even if some components are either exceedingly slightly volatile or exceedingly insoluble in the liquid phase. First we take the simplest case.

Raoult's law is

$$P = P°X \tag{10-12}$$

$P°$ is the vapor pressure of a pure substance, P is the vapor pressure of the same substance over a solution in which its mole fraction is X, both pressures being measured at the same temperature. In relatively rare cases Raoult's law applies to all components of a solution with good approximation. In general, Raoult's law applies to the vapor pressure of a component with increasingly good approximation as the mole fraction of that component approaches unity.

---

[2] The concept of the standard state is useful chiefly for identifying such quantities as the standard enthalpy, the standard heat capacity, the standard chemical potential, the standard cell potential, etc. As we have said, when the standard state is implicitly defined in ways such as we have used above, it is an idealized state. It is a state in which the substance would exist if it could be made to behave ideally from the reference state to the state in which its activity would become unity. Usually, as illustrated by Problem 10-4, a real state can be found in which $a = 1$. However, difficulties arise if one treats this real state as the standard state. For a brief discussion of this problem see *Chemical Thermodynamics* by Frederick T. Wall, Second edition, pp. 332–334, W. H. Freeman and Co., 1958.

Henry's law is

$$P_i = k_i X_i \tag{10-13}$$

P_i is the vapor pressure of a component over a solution in which its mole fraction is X_i; k_i varies with temperature, solvent, and solute, but is independent of X_i in the range in which this "law" holds. Conformance to Henry's law increases as $X_i \to 0$.[3]

We conclude, then, that in a dilute solution the solutes will usually conform closely to Henry's law and the solvent to Raoult's law.

PROBLEM 10-6   A solution in which the solutes obey Henry's law and the solvent obeys Raoult's law may be called an ideal dilute solution. By reference to 9-13 show that at any given temperature in an ideal dilute solution in which a general chemical reaction of the Equation 9-1 can occur, and if the vapors are ideal gases, there is a function of the mole fractions which will be a constant at chemical equilibrium regardless of the individual values of the X's; show that K_x, the equilibrium constant in solution is

$$K_x = \frac{X_H^h X_I^i \cdots}{X_B^b X_C^c \cdots} \tag{10-14}$$

■■

PROBLEM 10-7   Show that if X_i approaches zero, the mole fraction X_i, the molality m_i and the molarity C_i all become proportional to each other. Hence, show that Henry's law can be written as

$$P_i = k_i' m_i \tag{10-15}$$

or

$$P_i = k_i'' C_i \tag{10-16}$$

and that in very dilute solutions for which K_x is a constant, similar functions of the C's and m's are also constants, namely,

$$K_C = \frac{C_H^h C_I^i \cdots}{C_B^b C_C^c \cdots} \tag{10-17}$$

$$K_m = \frac{m_H^h m_I^i \cdots}{m_B^b m_C^c \cdots} \tag{10-18}$$

Equations 10-14, 10-17, and 10-18 represent approximations often applied to equilibria in dilute solutions.

■■

[3] It must be borne in mind that X_i refers to the same species as that to which P_i refers. Problems arise in applying Henry's law to cases such as those of acetic or hydrochloric acids. In the case of acetic acid the solute exists largely as a monomer in dilute aqueous solution, largely as a dimer in the vapor at ordinary pressures. In the case of HCl, the solute exists as ions in solution and as molecules in the vapor. Methods are developed later for handling such cases.

ACTIVITIES AND ACTIVITY COEFFICIENTS: CHEMICAL EQUILIBRIUM

We are now prepared to consider more specifically some conventional definitions of standard states of components of solutions. We can use nearly the same procedure we followed in defining the standard state of gases. First, for any particular constituent of the solution, B for example, we define a ratio a'_B/a''_B by

$$\mu'_B - \mu''_B = RT \ln \frac{a'_B}{a''_B} \qquad (10\text{-}19)$$

where the single and double primes refer to solutions of different concentrations at the temperature T.

∷

PROBLEM 10-8 By considering a liquid solution in equilibrium with a vapor phase and by reference to 10-5, 10-13, 10-15, and 10-16 show that a'_B/a''_B can be considered the ratio of the mole fractions, the molarities, or the molalities of B in the two solutions, but corrected to take account of nonideality.

∷

In order to be able to assign numbers to the a's we have to do something like what we did in the gas case in asserting that $f/P \to 0$ as $P \to 0$. This is equivalent to making an explicit or an implicit definition of the standard state because the standard state is always defined as one in which a, called the *activity*, is equal to 1. For the solvent in a dilute solution the definition is usually explicit: the standard state is the pure solvent. For solutes which do not dissociate[4] the usual convention is to write

$$a/C \to 1 \quad \text{as} \quad C \to 0 \qquad (10\text{-}20)$$

or

$$a/m \to 1 \quad \text{as} \quad m \to 0 \qquad (10\text{-}21)$$

However, one may also put the activity of solutes on a mole fraction basis and write

$$a/X \to 1 \quad \text{as} \quad X \to 0 \qquad (10\text{-}22)[5]$$

[4] The important case of solutes which dissociate is that of ionic substances. This case is considered later.

[5] Except for dilute solutions, the words solvent and solute are of indefinite meaning. In aqueous solutions water is often considered the solvent whether or not the solution is dilute. In other cases often no distinction is made, and the standard state of each component is taken as the pure component, that is, one writes

$$a/X \to 1 \quad \text{as} \quad X \to 1$$

for each component.

For a homogeneous solution reaction, and by the reasoning used in solving Problem 10-5, we can write rigorously

$$K = \frac{a_H{}^h a_I{}^i}{a_B{}^b a_C{}^c} \tag{10-23}$$

in which K is a constant at a given temperature for a given reaction. Moreover,

$$\Delta G^\circ = -RT \ln K \tag{10-24}$$

where the standard states to which ΔG° refers are those upon which the a's which appear in the expression for K are based. Activities are sometimes treated as dimensionless: they can be defined as the ratio of the fugacity in a nonstandard state to that in a standard state. However, it is often advantageous to think, speak, and write of activities in the units of concentration in terms of which they are defined. To do so directly implies the standard state being used and lessens the chance of error in calculations.

PROBLEM 10-9   Inasmuch as K in 10-24 is the argument of a logarithm it might appear that it should be dimensionless. However, when arguments of logarithms are factored and the logarithm is made into the sum of several logarithms, the factors may have dimensions even though the original argument was dimensionless. As an example of such a sum of logarithms recall 8-26, the Sackur–Tetrode equation. Using 10-19 and any other useful relations show that for a reaction which starts with a stoichiometric mixture of only reactants B and C at activities a_B and a_C and which goes to completion so that the final state is only the products H and I at the activities a_H and a_I

$$\Delta G = -RT \ln K + RT \ln \frac{a_H{}^h a_I{}^i}{a_B{}^b a_C{}^c} \tag{10-25}$$

It should be evident that if the two terms on the right side of 10-24 are combined into a single logarithm, the argument will be dimensionless, and that 10-24 is just a special case of 10-25 for which the numerical value of the argument of the second logarithm has become 1 but retains the dimensions complementary to the argument of the first logarithm.

::

In accord with the foregoing definitions, activities are equal to pressures or mole fractions or molalities or molarities in the limiting cases. Moreover, they may remain nearly equal to these things over considerable ranges of concentration. Even at concentrations where the activities depart more than a little from the corresponding directly measurable quantity, small variations in concentration may produce almost proportional variations in the activity. For these and other reasons it is convenient to define the *activity coefficient* γ^6 by expressions such as

$$a = \gamma P \qquad (=f) \tag{10-26}$$

$$a = \gamma X \tag{10-27}$$

$$a = \gamma C \tag{10-28}$$

$$a = \gamma m \tag{10-29}$$

EXAMPLES OF CHEMICAL POTENTIOMETRY

In deriving our equilibrium constant expressions 10-14, 10-17, 10-18, and 10-23 we made use of a heterogeneous system, that is, one having liquid and vapor phases each containing some of all the reacting substances. In many heterogeneous chemical reactions one or more of the phases may be considered pure substances. No complication, but rather a simplification is introduced in writing equilibrium constant expressions in such cases. Consider, for example, the reaction

$$4Ag + O_2 \rightarrow 2Ag_2O \tag{10-30}$$

In accord with 10-23 we would write

$$K = \frac{a_{Ag_2O}^2}{a_{Ag}^4 a_{O_2}} \tag{10-31}$$

But if we take, as we usually would, the standard states of the Ag and the Ag_2O as the pure substances under 1 atm pressure we can write to good approximation

$$K = \frac{1}{a_{O_2}} = \frac{1}{f_{O_2}} \tag{10-32}$$

A further simplification results if we can use the ideal gas approximation and write

$$K = \frac{1}{P_{O_2}} \tag{10-33}$$

Usually a treatment like that leading to 10-33 would be satisfactory. However, if the necessary data were available, the approximations would not have to be made. By considering the derivation it can be seen that 10-4 is generally applicable and is not restricted to any classes of systems such as ideal gases. By combining 10-4 with 10-19 we get

$$RT \ln \frac{a'}{a''} = \int_{P''}^{P'} V \, dP \tag{10-34}$$

---

[6] Some authors use different symbols for the different kinds of activity coefficients defined by 10-26, 10-27, 10-28, and 10-29.

Thus, with a knowledge of the equations of state of the substances involved in the reaction we can calculate the activities of these substances as functions of the pressure on the system.

PROBLEM 10-10   Suppose that reaction 10-30 were being studied under 100 atm pressure (the equilibrium temperature would be about 665 °K). Taking the densities of Ag and Ag_2O under these conditions as about 10 and 7 g cm^{-3} respectively, and making the approximation that these substances are not compressible, estimate the activities of Ag and Ag_2O at 665 °K and 100 atm.

Answer: 1.02 and 1.06

::

PROBLEM 10-11   Hilsenrath, Beckett, Benedict, Fano, Hoge, Masi, Nuttall, Touloukian, and Woolley[7] give as the virial equation of state for oxygen (and other gases)

$$PV = RT(1 + BP + CP^2 + DP^3) \tag{10-35}$$

where for oxygen at 900 °K they give

$$B = 0.3003 \times 10^{-3} \text{ atm}^{-1}$$

$$C = 0.15 \times 10^{-6} \text{ atm}^{-2}$$

$$D = 0.3591 \times 10^{-9} \text{ atm}^{-3}$$

On the basis of 10-9 and 10-25 and the above data estimate the activity (fugacity) and activity coefficient (fugacity coefficient) for the oxygen in the equilibrium system described in Problem 10-10, namely for O_2 at 665 °K, 100 atm.

Answer: $a = 103.1$ atm, $\gamma = 1.031$

::

PROBLEM 10-12   If one could increase the pressure on a condensed but volatile substance while keeping the condensed substance in equilibrium with its vapor, the vapor pressure should increase. To the extent that the vapor can be treated as an ideal gas, a valid equation is

$$RT \ln \frac{P'}{P} = \int_P^{P''} V \, dP \tag{10-36}$$

The left-hand member of this equation is the increase in chemical potential of the vapor in going from the normal equilibrium vapor pressure P to the vapor pressure P' over the condensed phase, on which the pressure has been increased from P to P''. The right-hand member of 10-36 is the increase in chemical potential for the condensed phase in going from P to P''. The validity of 10-36 depends upon the fact that for the system to remain in equilibrium,

[7] Above named authors, "Tables of Thermal Properties of Gases," National Bureau of Standards Circular 564 (1955).

the two increases in chemical potential must be equal. Except for the relatively small extra pressure in very small droplets caused by surface tension, it is difficult to devise unambiguous experiments to test 10-36. But to get an idea of the order of magnitude of the effect, calculate the fractional increase in water vapor pressure when the pressure on the liquid is raised from 1 to 2 atm at 300 °K. Show that a simplified form of 10-36 applicable to cases such as the above is

$$\frac{\Delta P_v}{\Delta P_1} = \frac{V_1}{V_v} \tag{10-37}$$

where V_1 and V_v are molal volumes of liquid and vapor.

∷

PROBLEM 10-13 The standard chemical potential (molal free energy) of diamond is 0.685 kcal mole^{-1} greater than that of graphite at 25 °C. The densities are for diamond 3.51 and for graphite 2.25 g ml^{-1}. Calculate the pressure which must be applied to a graphite-diamond mixture so that the two forms of carbon will have the same chemical potential. Show that the general relation is approximately

$$\mu_B^\circ - \mu_A^\circ = -M\left(\frac{1}{\rho_B} - \frac{1}{\rho_A}\right)(P - 1) \tag{10-38}$$

where M is the gram molecular (atomic) weight, P is the equilibrium pressure in atmospheres and the ρ's are the densities in grams ml^{-1}. The μ's would be in ml atm mole^{-1}.

Assuming that graphite is not less compressible than diamond, will the actual equilibrium pressure be more or less than that calculated by 10-38 using ρ's measured at 1 atm? ΔH_f° for diamond is 453 cal mole^{-1}. Will the equilibrium pressure be more or less at higher temperatures?

∷

PROBLEM 10-14 Use the data in Problem 10-2 to calculate the activity of the water in the specified solution. The proper method of calculation of the corresponding activity coefficient, γ, from 10-26 is ambiguous until a decision has been made as to how to calculate the mole fraction of H_2O. One possible decision would be to neglect ionization and formally define

$$X_{H_2O} = \frac{n_{H_2O}}{n_{H_2O} + n_{H_2SO_4}}$$

in which $n_{H_2SO_4}$ is the weight of H_2SO_4 divided by the formula-weight of H_2SO_4. Adopting this convention calculate γ_{H_2O} for the specified solution. The density of this solution is 1.25.
Answer: $\gamma_{H_2O} = 0.77$

∷

PROBLEM 10-15 One could take a different formal basis for the calculation of γ by arbitrarily calling the number of moles of solute three times the number of formula-weights of H_2SO_4. Such a treatment would presumably be based on the equation $H_2SO_4 \rightarrow 2H^+ + SO_4^=$ but without consideration of the incompleteness of ionization. On the basis of such a treatment show that for the solution under consideration $\gamma_{H_2O} = 0.90$. Neither this answer nor the answer to Problem 10-14 is incorrect, but these problems indicate that although in certain kinds of cases the activity co-efficient is a very useful variable, in others it is not.

::

The calculations of Problems 10-2, 10-14, and 10-15 were based on the use of a vapor-phase chemical potentiometer. That is, we read the chemical potential of the water in the solution by measuring the pressure of the water vapor in equilibrium with the solution. Once chemical potentials or activities of components of solutions are known, these solutions can be used as chemical potentiometers with respect to other systems. An illustration of this procedure is the work of Masterton and Gendrano[8] on solutions of

Table 10-1   The Concentration in Moles per Liter and the Activity of Water in 1,2-Dichloroethane in Equilibrium with Aqueous $CaCl_2$ Solutions at 25 °C.

	$CaCl_2$	
C_W	m	a_W
0.0696		1.000
0.0637	1.073	0.940
0.0581	2.025	0.857
0.0500	3.004	0.745
0.0411	3.959	0.619
0.0323	4.945	0.488
0.0286	5.494	0.422

water in organic solvents. These authors equilibrated aqueous $CaCl_2$ solutions with several organic solvents and then analyzed the organic phase to determine the concentration of water in it. Some of the data are reproduced in Table 10-1. The values given for a_W are the values obtained by other investigators from vapor pressure or related measurements on aqueous $CaCl_2$ solutions.[9] The authors analyzed their data to indicate that dissolved in some organic solvents water is very largely monomeric (benzene) but that in others (dichloroethane) it is appreciably dimerized.

[8] W. L. Masterton and M. C. Gendrano, "Henry's Law Studies of Solutions of Water in Organic Solvents," *J. Phys. Chem.*, **70**, 2895 (1966).
[9] See M. Lietzke and R. Stoughton, *J. Phys. Chem.*, **66**, 508 (1962) and R. Stokes, *Trans. Faraday Soc.*, **41**, 637 (1945).

PROBLEM 10-16   Taking the vapor pressure of pure water at 25 °C as 23.76 torr calculate Henry's law "constants" as defined by 10-16 for each of the solutions of water in dichlorethane listed in Table 10-1.

::

PROBLEM 10-17   The activities shown in Table 10-1 are based on the convention that pure water is the standard state, i.e., that $a \to 1$ as $X \to 1$. Using this standard state and 10-27 as the definition of γ, calculate the γ's for the H_2O in the dichloroethane solutions tabulated in Table 10-1.

::

PROBLEM 10-18   Alternatively, one might treat water as a solute and let $a/C \to 1$ as $C \to 0$ (10-20). The difficulty of treating these data by this convention is the difficulty of extrapolating to zero concentration. Correct relative values would result from taking any other working definition of standard state. If our working implicit definition of the standard state is $a = C$ when $C = 0.0323$, show that the values of γ for the concentrations shown in Table 10-1 (based on 10-27) are: 0.915, 0.977, 0.977, 0.987, 0.997, 1.000, 0.978. (To solve this problem observe that defining Equation 10-18 requires that the value of the activity corresponding to a substance in a specified state may have different values for different choices of standard state, but that the ratios of the a's corresponding to any specified pair of states must be the same for one choice of standard state as for any other choice.)

TABULATION OF STANDARD THERMODYNAMIC PROPERTIES OF SUBSTANCES

The most common way in which the chemical potentials of substances in their standard states are specified is in terms of standard free energies (Gibbs energies) of formation. These are defined on a basis entirely analogous to the way in which standard enthalpies of formation were defined in Chapter 1. The definition is indicated by Figure 10-1. On the basis of this definition the standard Gibbs energy of formation of elements in their standard states is

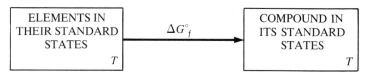

FIGURE 10-1   Definition of the standard Gibbs energy of formation of a compound.

taken as zero at all temperatures. This procedure might seem inconsistent with

$$\left(\frac{\partial G}{\partial T}\right)_P = -S \tag{5-25}$$

But there is no inconsistency. The standard states at different temperatures are different states, and by assigning to ΔG_f° the value zero for elements in all of these states we are just shifting our base line—which is in any case arbitrary—each time we change the temperature.

An alternative way of specifying standard chemical potentials is in terms of the free energy function

$$\frac{G^\circ - H_0^\circ}{T} \quad \begin{bmatrix} \text{Standard} \\ \text{free energy} \\ \text{function} \end{bmatrix} \tag{10-39}$$

This function, as we saw, has an especially simple relation to the partition function

$$\left(\frac{G - H_0}{T}\right)_{int} = -R \ln Q \tag{8-44}$$

An obviously closely related function is

$$\frac{G^\circ - H_{298.15}^\circ}{T} \quad \begin{bmatrix} \text{Standard} \\ \text{free energy} \\ \text{function based} \\ \text{on } H_{298.15}^\circ \end{bmatrix} \tag{10-40}$$

The advantages of the use of these functions and examples of their use are explained by Pitzer and Brewer, and the reader is referred to their book,[10] Chapter 15 for the discussion and Appendix 7 for extensive tables from which a small selection of values are reproduced here in Tables 10-3A, 10-3B, and 10-3C. To facilitate finding them from time to time, Tables 10-2 and 10-3 are placed at the end of the book as Appendixes A and B.

PROBLEM 10-19   It can readily be seen that if the sum of the standard free energy functions for reactants is subtracted from the sum of those for the products 10-41 results.

$$\sum \left(\frac{G^\circ - H_{298}^\circ}{T}\right)_{\text{Products}} - \sum \left(\frac{G^\circ - H_{298}^\circ}{T}\right)_{\text{Reactants}} = \frac{\Delta G^\circ - \Delta H_{298}^\circ}{T} \tag{10-41}$$

where ΔG° is the ΔG° for the reaction at the temperature corresponding to the values used for the free energy function and ΔH_{298}° is ΔH° for the reaction at 298 °K. (An entirely analogous equation could be written involving the

[10] G. N. Lewis and M. Randall, *Thermodynamics*, revised by K. S. Pitzer and L. Brewer, McGraw-Hill, 1961.

free energy functions defined by 10-39 based on ΔH_0°.) Using 10-41 and data from Table 10-3 calculate ΔG_f° for NaCl at 25 °C and note that the result agrees with the value shown in Table 10-2. (See Appendixes A and B.)

##

Measurement of the electrical potentials of reversible electro-chemical cells has been a fruitful method of determining the chemical potentials of substances involved in the cell reactions. The procedure is direct in the case of formation cells, that is, of cells in which the cell reaction is the formation of a compound from its elements. The simplicity in that case arises because the standard chemical potentials of the reactants (elements) is zero. The calculation is not complicated much, however, if the compound of interest is produced from other compounds whose chemical potentials are known.

As an example we consider the cell

$$\text{Pt, H}_2(\text{atm}) \mid \text{HCl}(m) \mid \text{AgCl(c), Ag(c)} \tag{10-42}$$

in which the cell reaction per faraday is

$$\tfrac{1}{2}\text{H}_2 + \text{AgCl} \to \text{Ag} + \text{HCl} \tag{10-43}$$

Here (m) means a solution which is m molal in HCl and (c) means the crystalline state at 1 atm which is the standard state for the solid substances. This cell has been studied extensively by Noyes and Ellis,[11] by Linhart,[12] by Harned and Ehlers[13] and others. Some of the data from the experiments of Harned and Ehlers are reproduced in Table 10-4.

Table 10-4 Potentials (volt) of the Cell Pt, H_2(1 atm) | HCl(m) | AgCl(c), Ag(c) at 25 °C as a Function of the Molal Concentration of the HCl.

m	ϵ	m	ϵ
0.00356	0.51527	0.04986	0.38582
0.00449	0.50384	0.05391	0.38222
0.01002	0.46376	0.09642	0.35393
0.01031	0.46228	0.2030	0.31774

The relation between the cell potential and the chemical potentials is obtained from

$$dG = -\mathscr{F}\varepsilon \, dn \tag{5-19}$$

and

$$\widetilde{\Delta\mu} = \frac{dG}{d\xi} = -\text{b}\mu_\text{B} - \text{c}\mu_\text{C} - \cdots + \text{h}\mu_\text{H} + \text{i}\mu_\text{I} + \cdots \tag{10-44}$$

[11] A. A. Noyes and Ellis, *J. Am. Chem. Soc.*, **39**, 2532 (1917).
[12] G. A. Linhart, *J. Am. Chem. Soc.* **41**, 1175 (1919).
[13] H. S. Harned and R. W. Ehlers, *J. Am. Chem. Soc.*, **54**, 1350 (1932).

10-44 is the same as 9-6 except that now we are concerned with a system not at chemical equilibrium (so that $\Delta\tilde{\mu}$ is not zero) and one which can react with its surroundings electrically as well as through T and PV interactions (so that $\Delta\tilde{\mu}$ need not be less than zero). The dn in 5-19 is the number of faradays passed through the cell. For the case that the cell reaction is written to correspond to one faraday, $dn = d\xi$. For other cases $dn = \tilde{n}\,d\xi$ where $\tilde{n}$ is the number of faradays corresponding to the cell reaction as written. We can then combine 5-19 and 10-44 to get the equation applicable to a reversible electrochemical cell at constant T and P.

$$\widetilde{\Delta\mu} = -\tilde{n}\mathscr{F}\varepsilon \qquad (10\text{-}45)$$

Thus, if all of the μ's except one (of the right-hand member of 10-44) are known, that one can be calculated from a measurement of ε.

PROBLEM 10-20   In the cell 10-42 the AgCl and the Ag are in their standard states. The H_2 is nearly in its standard state. If it were in its standard state, its chemical potential (Gibbs energy of formation) could be taken as zero. For the most precise calculations one would get its actual chemical potential in the cell by the use of an appropriate equation of state with 10-5 and 10-10. Making the approximation that the chemical potential of H_2 in the cell is zero and using the value of -26.24 kcal mole^{-1} for the standard free energy (Gibbs energy) of formation of AgCl at 25 °C, calculate the chemical potential of HCl at 25 °C in (a) a 0.203 m solution and (b) in a 0.0103 m solution.
Answer: (a) -33.57, (b) -36.90 kcal mole^{-1}

∷

PROBLEM 10-21   Taking the standard chemical potential of HCl gas at 25° as -22.77 kcal mole^{-1}, estimate the fugacities (activities based on the ideal gas standard state) of HCl in the two solutions specified in Problem 10-20. Note that these will be very nearly equal to the HCl vapor pressures over these solutions and also that they are roughly proportional to the square of the corresponding molality.
Answer: (a) 1.2×10^{-8} atm, (b) 4.0×10^{-11} atm (The number of figures given in the answers to Problems 10-20 and 10-21 is not meant to represent the accuracy with which the calculation could be made on the basis of the data quoted.)

∷

Another electrochemical example will again illustrate the use of an intermediate chemical potentiometer (isotensiscope). The $CaCl_2$ solutions were used in this way in the work, previously discussed, of Masterton and Gendrano. We shall cast our problem as being the determination of the chemical potential of NaCl at specified concentrations in aqueous solution.

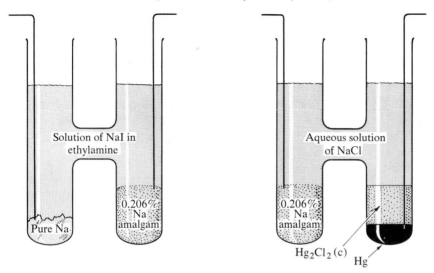

FIGURE 10-2

We cannot use exactly the same method as the one used in the HCl case because unlike hydrogen, sodium is not stable in contact with aqueous solutions. Lewis and Kraus[14] ingeniously solved this problem by what amounts to the determination of the chemical potential of a dilute sodium amalgam and then the use of this dilute sodium amalgam, which is sufficiently stable toward water, in an aqueous NaCl cell. Figure 10-2 shows the cells on the basis of which the analysis and calculation can most easily be made.

Taking the left electrode of each cell as the anode, the cell reactions per faraday would be

A. $\quad Na(c) \rightarrow Na(0.206\% \text{ amalgam})$ $\qquad$ (10-46)

B. $\quad Na(0.206\% \text{ amalgam}) + \frac{1}{2}Hg_2Cl_2(c) \rightarrow Hg(l) + NaCl(m)$ $\quad$ (10-47)

The net reaction per faraday if current were passed through the two cells in series would be

$$Na(c) + \tfrac{1}{2}Hg_2Cl_2(c) \rightarrow Hg(l) + NaCl(m) \qquad (10\text{-}48)$$

Lewis and Kraus found for the electrical potential of cell A at 25 °C 0.8453 volt. Lewis calculated from the data of Allmand and Polack[15] that for cell B when it contained 1.022 m NaCl, $\varepsilon = 2.1582$ volt at 25 °C.

[14] G. N. Lewis and C. A. Kraus, *J. Am. Chem. Soc.*, **32**, 1459 (1910).

[15] A. J. Allmand and W. G. Polack, *J. Chem. Soc.*, **115** 1020 (1919). In their original work Lewis and Kraus studied a cell like cell B but which required more complicated calculations. For a comparison of the two researches, in addition to the original papers see *Thermodynamics* by Lewis and Randall, First edition, McGraw-Hill, 1923, pp. 352, 414–417.

PROBLEM 10-22   Calculate the chemical potential, the activity and the activity coefficient of sodium in a 0.206% amalgam at 25 °C. Take pure sodium as the standard state and use 10-26 as the definition of γ.
Answer: $\mu = -19.50$ kcal mole^{-1}; $a = 5 \times 10^{-15}$; $\gamma = 3 \times 10^{-13}$.

::

PROBLEM 10-23   Taking ΔG_f° for Hg_2Cl_2 as -50.350 kcal mole^{-1}, calculate the chemical potential of NaCl in a 1.022 m aqueous solution. Make use of the fact that the electrical potential corresponding to 10-48 will be the sum of the ε's of cells A and B.
Answer: -94.44 kcal mole^{-1}.

11

Ions

On the one hand, thermodynamics is a discipline that can extract a multitude of useful facts and make a great variety of reliable predictions from experimental data which do not include or require knowledge of the fine structure of its systems. On the other hand such knowledge, when available, and even if incomplete, often helps greatly in setting up rational and relatively simple thermodynamic formulas and notations. An important example of this fact is the thermodynamic treatment of substances which ionize in solution. This chapter is concerned chiefly with descriptions and examples of the use of the conventions which have been found useful and have been widely adopted for the thermodynamic treatment of systems containing dissolved ions. These conventions are then available to us so that later we can, to whatever extent we choose, include electrolytes in our study of some of the important aspects of the thermodynamics of solutions.

IONIZATION

In 1883 Arrhenius[1] proposed that when dissolved in water certain substances dissociate into positively and negatively charged molecules, *ions*. Thus, the ionization of HCl can be represented by

$$HCl \rightarrow H^+ + Cl^- \tag{11-1}$$

In itself this fact might not seem to call for any special thermodynamic treatment of such substances. For example, in accord with 10-23 we might write

$$K = \frac{a_{H^+} a_{Cl^-}}{a_{HCl}} \tag{11-2}$$

[1] Clausius had made such a suggestion in 1857 but apparently did not pursue the idea and develop the theory, which Arrhenius did.

Among the evidences that HCl does largely dissociate is the fact indicated by the solution to problem 10-27 that with respect to the two given concentrations, the activity of HCl is roughly proportional to the square of its concentration.

Two great difficulties exist with respect to the use of 11-2. They both have to do with the evaluation of the activities of the ions, a_{H^+} and a_{Cl^-}. The first involves the question of determining the concentrations of the HCl, the H^+ and the Cl^-. It is clear that in an aqueous solution containing only HCl, $C_{H^+} = C_{Cl^-}$ and $C_{HCl} = C - C_{H^+}$. Thus if the degree of dissociation could be determined, the C's could be evaluated. For substances such as acetic acid which are only slightly dissociated at moderate concentrations, meaningful values of the degree of dissociation can be determined, for example by conductance methods. For substances such as HCl, however, various kinds of evidence lead to the conclusion that these substances can best be treated, for many purposes, as substantially completely dissociated at all moderate concentrations and that it is often a very difficult and unnecessary task to seek an experimental value of the K in 11-2.[2] Rather, an arbitrary assignment of the value of one to K is made as part of the implicit definition of the standard state of the solute. This procedure constitutes the solution to, or more exactly, the circumvention of the first major difficulty in the thermodynamic treatment of systems like aqueous HCl.

The second difficulty arises from the fact that because of their charge it is impossible to transfer or change the state of a single species of ions in thermodynamic (macroscopic) amounts without simultaneously involving ions of the opposite charge. Therefore the methods of chemical potentiometry discussed in the preceding chapter are inapplicable, and no generally accepted method has been proposed for determining the activity or the activity coefficient of a single species of ions.[3] Although it is sometimes convenient to use the symbols for individual ion activities, the impossibility of their unambiguous evaluation by experimental procedures so far proposed is accepted, and it is recognized that only products of ion activities such as appear in 11-2, or mean ion activities and activity coefficients can be thermodynamically measured.[4]

---

[2] This treatment of strong electrolytes as being completely dissociated at all moderate concentrations is an example of a fact that has been mentioned before. It is the fact that various self-consistent ways can be set up for formulating the implications of thermodynamic laws. A somewhat different treatment will be found in C. B. Monk's excellent book *Electrolytic Dissociation*, Academic Press, 1961, more than half of which is concerned with the question of incomplete dissociation of relatively strong electrolytes.

[3] There are those, however, who take the position that meaningful values can be assigned to the activity coefficient of a single ion species. See, for example, Henry S. Frank, "Single-Ion Activities and Ion-Solvent Interaction in Dilute Aqueous Solutions," *J. Phys. Chem.*, **67**, 1554 (1963).

[4] One can also measure ratios of activities of like-sign ions. For example, from measurements on the cell Cu, CuO/K_2SO_4(aq), KOH(aq)/$PbSO_4$, Pb one could calculate $a_{SO_4^-}/a^2_{OH^-}$.

ACTIVITIES AND ACTIVITY COEFFICIENTS OF ELECTROLYTES

Among the monumental contributions of G. N. Lewis to the science of chemical thermodynamics can be counted his development of experimental and theoretical methods of dealing with electrolytes. Probably a decisive factor in the success of these developments was his recognition as early as 1909 of evidence suggesting that strong electrolytes are virtually completely dissociated up to concentrations of the order of 1 normal. It was advantageous, therefore, to handle the thermodynamics of strong and weak electrolytes somewhat differently. We have already alluded to the chief difference. If we take HAc as the typical weak electrolyte we can exemplify the difference in convention between strong and weak electrolytes by writing

$$HAc \rightarrow H^+ + Ac^- \tag{11-3}$$

$$K = \frac{a_{H^+} a_{Ac^-}}{a_{HAc}} \tag{11-4}$$

or

$$a_{HAc} = \frac{1}{K} a_{H^+} a_{Ac^-} \tag{11-5}$$

But instead of 11-2, for the strong electrolyte HCl we write

$$1 = \frac{a_{H^+} a_{Cl^-}}{a_{HCl}} \tag{11-6}$$

or

$$a_{HCl} = a_{H^+} a_{Cl^-} \tag{11-7}$$

Inasmuch as we cannot determine the separate values of a_{H^+} and a_{Cl^-} we can define a mean ion activity, $a_\pm$

$$a_\pm^2 = a_{H^+} a_{Cl^-} \tag{11-8}$$

One next defines ionic activity coefficients by equations such as

$$a_{H^+} = \gamma_{Cl_{H^+}} m_{H^+} \tag{11-9}$$

$$a_{Cl^-} = \gamma_{Cl^-} m_{Cl^-} \tag{11-10}$$

$$a_\pm^2 = \gamma_\pm^2 m_{H^+} m_{Cl^-} \tag{11-11}[5]$$

We then complete our specification of the means for calculating values of the γ's and of a and $a_\pm$ by stating that $\gamma_\pm \rightarrow 1$ as the total ion concentration $\rightarrow 0$. It should be noted that it is not a sufficient condition for $\gamma_\pm \rightarrow 1$ that the

[5] Similar activity coefficients can be defined on a molarity or mole fraction basis.

concentration of the ions of the electrolyte of interest should approach zero. We discuss later the fact that the value of $\gamma_\pm$ depends upon the concentration of all the ions present in the solution.

The defining equations 11-8, 11-9, 11-10, 11-11, and the statement about $\gamma_\pm \to 1$ as the ion concentration $\to 0$ are applied to both strong and weak electrolytes. Accordingly we can put 11-5 and 11-7 in the forms

$$a_{\text{HAc}} = \frac{1}{K} \gamma_\pm^2 m_{\text{H}^+} m_{\text{Ac}^-} \tag{11-12}$$

and

$$a_{\text{HCl}} = \gamma_\pm^2 m_{\text{H}^+} m_{\text{Cl}^-} \tag{11-13}$$

One way of interpreting the difference between 11-12 and 11-13 (or between 11-5 and 11-7) is to say that it is possible and convenient to specify acceptable ways of independently evaluating a_{HAc} and $a_{\pm \text{H}^+\text{Ac}^-}$ and hence to calculate K's which turn out to be nearly independent of the concentrations, whereas with HCl and other strong electrolytes this is not the case.

There is another way of interpreting the difference between the weak and strong electrolyte equations. In both cases, a standard state of the ions is implicitly defined by the following relations together with the usual convention that in the standard state $a_\pm = 1$.

$$\mu' - \mu'' = 2RT \ln \frac{a'_\pm}{a''_\pm} \tag{11-14}$$

and

$$\frac{a_\pm^2}{m_+ m_-} \to 1 \text{ as total ion concentration} \to 0 \tag{11-15}$$

In the case of weak electrolytes the implied standard state of the unionized substance can conveniently be independently specified by similar equations

$$\mu' - \mu'' = RT \ln \frac{a'}{a''} \tag{10-19}$$

and

$$a/m \to 1 \quad \text{as} \quad m \to 0 \tag{10-21}$$

Although the extrapolation of the properties of the unionized species to $m \to 0$ would seem to give trouble because of the increasing extent of dissociation, in some cases a good approximation is to neglect the dissociation even at concentrations so small that the activity coefficient of the undissociated substance is substantially 1, in other cases the extent of dissociation can be allowed for in the calculation. Having then implicitly defined standard states for both unionized and ionized solutes it is possible with suitable experimental data to calculate a useful thermodynamic equilibrium constant from 11-4. The difference between this situation for weak electrolytes and that for strong electrolytes is that with strong electrolytes the standard state of the undissociated solute is not defined independently, but is implied by the following conventions and thermodynamic relations (exemplified by the case of HCl):

$$a_{\mathrm{HCl}} = \gamma_{\pm}^2 m_{\mathrm{H}^+} m_{\mathrm{Cl}^-} \tag{11-13}$$

$$\gamma_{\pm} \to 1 \text{ as total ion concentration} \to 0 \tag{11-16}$$

$$\mu' - \mu'' = RT \ln \frac{a'_{\mathrm{HCl}}}{a''_{\mathrm{HCl}}} \tag{11-17}$$

In the standard state

$$a_{\mathrm{HCl}} = 1 \tag{11-18}$$

Before considering examples of the use of the foregoing equations we indicate how the conventions are extended to electrolytes other than the 1–1 type. For the record we shall set forth expressions for $a_{\pm}$ and $\gamma_{\pm}$ applicable to the general dissociation equation

$$A_x B_y \to xA^{z+} + yB^{z-} \tag{11-19}$$

We write

$$a_{\pm}^{(x+y)} = a_+^x a_-^y \tag{11-20}$$

$$\gamma_{\pm}^{(x+y)} = \gamma_+^x \gamma_-^y \tag{11-21}$$

and for the strong electrolyte

$$a_{A_x B_y} = a_{\pm}^{(x+y)} \tag{11-22}$$

and

$$a_{A_x B_y} = \gamma_{\pm}^{(x+y)} m_{A^{z+}}^x \, m_{B^{z-}}^y \tag{11-23}$$

But for clarifying illustration we write for BaCl$_2$

$$a_{\pm}^3 = a_{\mathrm{Ba}^{++}} a_{\mathrm{Cl}^-}^2 \tag{11-24}$$

$$\gamma_{\pm}^3 = \gamma_{\mathrm{Ba}^{++}} \gamma_{\mathrm{Cl}^-}^2 \tag{11-25}$$

$$a_{\mathrm{BaCl}_2} = a_{\pm}^3 = \gamma_{\pm}^3 m_{\mathrm{Ba}^{++}} m_{\mathrm{Cl}^-}^2 \tag{11-26}$$

Inasmuch as weak electrolytes dissociate in steps, the formulas for 1–1 type dissociations can be applied to the dissociation of weak electrolytes.

Although nonthermodynamic, data related to the conductivity of electrolyte solutions have contributed importantly to the development of the thermodynamics of electrolytes. A crude approximation, but one not entirely unsatisfactory in some cases and in certain concentration ranges, would be to treat the mobility of ions in an electric field as independent of the concentration. The conductance of any amount of solution of a particular electrolyte between parallel plates 1 cm apart would then be a measure of the number of ions of that electrolyte present. When the amount of solution is such as to contain one equivalent of the electrolyte the conductance of the system described is defined as the equivalent conductance Λ of the solution.

If one were dealing with a strong electrolyte, under the above-stated approximation one would expect Λ to be independent of concentration. That the approximation becomes rather poor at high concentrations is shown by the experimental fact that Λ can be well represented for strong 1–1 electrolytes up to concentrations of 1 molar or higher by the equation

$$\Lambda = \Lambda_0 - k\sqrt{C} \qquad (11\text{-}27)$$

Λ_0 is the equivalent conductance extrapolated to infinite dilution. (For higher valence type electrolytes conformance to 11-27 is found only at molarities much less than 1.) For example,[6] for HCl at 25 °C

$$\Lambda_{C=0.01} = 412.00$$

and

$$\Lambda_{C=0.1} = 391.32 \text{ ohm}^{-1} \text{ cm}^2 \text{ equiv}^{-1}$$

For weak electrolytes one would expect the kind of behavior described by 11-27 to be superimposed on a much larger effect, namely, the variation in extent of ionization with variations in concentration. For the dissociation of the weak electrolyte HA

$$\text{HA} \rightarrow \text{H}^+ + \text{A}^- \qquad (11\text{-}28)$$

$$K = \frac{a_{\text{H}^+}\, a_{\text{A}^-}}{a_{\text{HA}}}$$

If α represents the fractional dissociation and if we make the approximations that $\gamma_{\text{HA}} = \gamma_{\pm} = 1$, it follows that

$$\frac{C\alpha^2}{1-\alpha} = K \qquad (11\text{-}29)$$

But if we ignore the effect of 11-27, which will be a reasonably good approximation if both α and C are small,[7] we would conclude for a weak electrolyte that

$$\frac{\Lambda}{\Lambda_0} = \alpha \qquad (11\text{-}30)$$

and

$$\frac{C\Lambda}{\Lambda_0(\Lambda_0 - \Lambda)} = K \qquad (11\text{-}31)$$

11-31 is called Ostwald's dilution law.

[6] Values quoted from T. Shedlovsky, *J. Am. Chem. Soc.*, **54**, 1411 (1932).

[7] In considering the possible application of 11-27 to weak electrolytes it should be understood that the Λ and Λ_0 in that equation would refer to an equivalent *of the ions*, not to a total of one equivalent of the weak electrolyte. In 11-30 and 11-31 however, the Λ and the Λ_0 refer to a total of one equivalent of the weak electrolyte. To make the suggested correction, then, we would probably first calculate α using Λ_0 then we would replace Λ_0 11-30 and 11-31 by a corrected value calculated through 11-27 which would take account of the lessened mobility caused by the concentration of ions produced by the dissociation.

Table 11-1* Equivalent Conductances
of Acetic Acid Solutions.

C, moles l^{-1}	Λ, ohm^{-1} cm^2 equiv^{-1}
0.000494	68.22
0.000988	49.50
0.001976	35.67
0.003952	25.60
0.01581	13.03
0.03162	9.260
0.06323	6.561
0.2529	3.221
1.011	1.443
Λ_0	387.9

* Data from D.A. MacInnes and T. Shedlovsky, *J. Am. Chem. Soc.* **54**, 1429 (1932).

In order to test out 11-31 or to use it or refinements of it to calculate dissociation constants we must be able to evaluate Λ_0. That this cannot be done by extrapolation to zero concentration of the Λ's for weak electrolytes is evident from the data in Table 11-1. What can be done, however, is to deal with strong electrolytes and write

$$\Lambda_{0_{CH_3COOH}} = \Lambda_{0_{NaOOCCH_3}} + \Lambda_{0_{HCl}} - \Lambda_{0_{NaCl}} \qquad (11\text{-}32)$$

PROBLEM 11-1 Use the data in Table 11-1 together with the approximations represented by 11-31 to calculate a series of values for K. The values so calculated should be nearly constant at all the lower concentrations.

::

PROBLEM 11-2 Show that 11-31 is evidently inapplicable to the dissociation of HCl by substituting into 11-31 the values quoted above for Λ_0, $\Lambda_{0.01}$, and $\Lambda_{0.1}$.

::

THE IONIC STRENGTH PRINCIPLE

Lewis and Randall,[8] apparently largely empirically, discovered the fact that "in dilute solutions, the activity coefficient of a given strong electrolyte is the

[8] G. N. Lewis and M. Randall, *J. Am. Chem. Soc.*, **43**, 1112 (1921).

same in all solutions of the same ionic strength." They defined the ionic strength, for which we use the symbol I by

$$I = \tfrac{1}{2} \sum m_i z_i^2 \tag{11-33}$$

where the sum is taken over all the ions in the solution, m_i being the molality of and z_i the charge on the ith kind of ions.

::

PROBLEM 11-3   Calculate the ionic strength of (*a*) an x molal solutions of 1–1 strong electrolyte, (*b*) of a 1–2 type, (*c*) of a 2–2 type.
Answer: (*a*) x, (*b*) 3x, (*c*) 4x

::

PROBLEM 11-4   Calculate the ionic strength of a solution which is 0.01 m in $La_2(SO_4)_3$ *and* 0.01 m in K_2SO_4.
Answer: 0.18 m

::

If one were pretending to present anything like a complete historical survey of the development of the theory of electrolytes one would have to cite the papers of many who worked in the first two decades and the early part of the third decade of the twentieth century to whom much credit belongs. Certainly the names would include W. Sutherland, A. A. Noyes, N. Bjerrum, S. R. Milner, J. C. Gosh, J. N. Brønsted and G. N. Lewis, and M. Randall. That it remained for Debye and Hückel to give the impetus to very rapid and extensive development of the theory can be attributed partly to lack of widespread familiarity with the work of some of these earlier investigators but importantly, also, to the fact that by making approximations and assumptions Debye and Hückel were able to reduce the calculations to manageable form. Some of these approximations, such for example as that the dielectric constant as seen by an individual ion could be taken as the same as the bulk dielectric constant of the solvent, could appear quite crude. But one can find numbers of instances in science in which the making of even crude approximations has led to very valuable results which could not be achieved by more rigorous treatments because of the complexity of the problem.

DEBYE–HÜCKEL EQUATIONS

Debye and Hückel[9] proposed a relatively simple theory which not only accounts for the rule of Lewis and Randall quoted above, but which leads to equations which are in accord with the experimentally found quantitative relationship between the ionic strength and the activity coefficients of ions in

---

[9] P. Debye and E. Hückel, *Physik. Z.*, **24**, 185, 334 (1923), **25, 97** (1924).

dilute solutions. These equations have been extremely useful and they have been elaborated by a number of investigators.[10] As a working statement of the results of these treatments we present 11-34, an equation proposed by Guggenheim.[11] Using 0.2 for b, Davies[12] has found this equation to hold for aqueous solutions at 25 °C for 1–1 and 2–1 electrolytes to within about 2% for values of I up to 0.1.

$$\log \gamma_{\pm} = -0.50 \,|z_+z_-| \left(\frac{\sqrt{I}}{1 + \sqrt{I}} - bI\right) \tag{11-34}$$

Here $|z_+z_-|$ means the product of the charge numbers, taken as a positive number, of the anion and cation for which $\gamma_{\pm}$ is the mean ion activity coefficient. By variation of the constants, similar equations can be applied to other solvents and at other temperatures. The initial constant (0.50) is a function of the density and dielectric constant of the solvent and of the temperature. It does not change rapidly with temperature. It is evident that at great dilutions the $0.2I$ as well as the $\sqrt{I}$ in the denominator become negligible and the result can be simplified. Partly as an indication of the magnitude of the temperature effect we present this simplified form of the limiting law equation for aqueous solutions at 25° and 0 °C.

$$\text{At } 25° \quad \log \gamma_{\pm} = -0.51 \,|z_+z_-|\, \sqrt{I} \tag{11-35}$$

$$\text{At } 0° \quad \log \gamma_{\pm} = -0.49 \,|z_+z_-|\, \sqrt{I} \tag{11-36}$$

We shall not derive 11-34, 11-35, and 11-36, but before making use of them, we briefly summarize the basic factors which underlie the distinction between the behavior of ionized and unionized solutes and which are the major premises of the derivations of these equations. These factors are:

1. The force between charged particles varies inversely with the dielectric constant of the medium and inversely with the square of the distance between them; it is attractive for unlike and repulsive for like charges.

2. The concentrations of positively and negatively charged ions in a solution are such as to produce a net electrical neutrality.

3. The forces between uncharged molecules decrease much more rapidly with distance than do the coulombic forces between ions.

Among the implications of these principles relevant to the present discussion are the following. Inasmuch as departures from ideal behavior of solutes can be attributed to forces acting among the ions and molecules, at a

[10] For a summary of the methods of Debye and Hückel and reviews of the subsequent contributions of other investigators see, for example, C. B. Monk, *Electrolytic Dissociation*, Academic Press, 1961. These developments are also taken up to various extents and degrees of sophistication by many texts on physical chemistry, electrochemistry, and chemical thermodynamics.

[11] E. A. Guggenheim, *Phil. Mag.*, **19**, 588 (1935).

[12] C. W. Davies, *J. Chem. Soc.*, 2093 (1938).

given concentration because of the relatively long-range nature of coulombic forces, departures from ideality would be expected to be much greater for electrolytes than for nonelectrolytes, which we know to be the experimental fact. The principle of electrical neutrality, as we have already seen, prevents us from evaluating single ion activities. It is also the basis of Brønsted's[13] principle of specific ion interaction. This principle—an approximation—considers a particular ion and treats its interaction with its surroundings as that between the ion and a surrounding "ionic atmosphere" of equal and opposite charge. In terms of this approximation, the coulombic forces that have to be considered in developing a theory for departures of electrolytes from ideal behavior are only attractive ones. Finally, the inverse dependence of the coulombic forces upon the dielectric constant of the solvent explains what is usually found to be the case, namely that solubilities and dissociation constants of electrolytes are greater in solvents of high than in ones of low dielectric constant.

The conformance of actual systems to 11-34 can be tested in many ways. For example, let us consider the solubility of the slightly soluble salt $Ba(IO_3)_2$. We write for the solubility product constant

$$K = a_{Ba^{++}} a_{IO_3^-}^2 \qquad (11\text{-}37)$$

We rewrite it

$$K = \gamma_{\pm}^3 m_{Ba^{++}} m_{IO_3^-}^2 = 4\gamma_{\pm}^3 m^3 \qquad (11\text{-}38)[14]$$

Monk gives the data we include in Table 11-2. C_1 is the molar concentration of $Ba(IO_3)_2$ in solutions saturated with $Ba(IO_3)_2$ at 25 °C and containing concentrations C_2 of KCl. The values of $\gamma_{\pm}$ shown in Table 11-2 are those he calculated from an equation equivalent to 11-38 (by substituting the C_1's for m).

Table 11-2 Mean Activity Coefficients of $Ba(IO_3)_2$
in KCl Solutions Calculated from the Measured
Solubilities.*

C_2	0.0	0.001	0.0035	0.005
$10^4 \times C_1$	8.10	8.40	8.59	8.74
$\gamma_{\pm}$	0.8958	0.8772	0.8448	0.8303
C_2	0.0075	0.010	0.020	0.050
$10^4 \times C_1$	8.99	9.18	9.85	11.17
$\gamma_{\pm}$	0.8069	0.7903	0.7365	0.6495

* C. B. Monk, *Electrolytic Dissociation*, Academic Press, 1961.

[13] N. Brønsted, *J. Am. Chem. Soc.*, **44**, 877 (1922).
[14] At ordinary temperatures and low concentrations the difference between the value of the molarity and the value of the molality in aqueous solutions is small, and we usually neglect it. The conversions among molality, molarity, and mole fraction are simple in theory but they are a bit tedious algebraically and in the case of molarity require a knowledge of the density of the solution.

PROBLEM 11-5 Calculate the ionic strengths of the solutions shown in Table 11-2. Calculate the values of $\gamma_\pm$ given by 11-34 (using $b = 0.2$) and by 11-35 and compare these calculations with each other and with the experimental values shown in Table 11-2.

::

We can illustrate the utility of Debye–Hückel equations in extrapolating the behavior of electrolyte solutions to the zero concentration reference state by considering the cell 10-42. We have not previously discussed standard cell potentials and the form of a general equation relating the cell potential to the activities of the cell reactants and products. The standard cell potential ε° is defined as the reversible potential when cell reactants and products are all in their standard states. We can write

$$-\widetilde{\Delta\mu} = \tilde{n}\mathscr{F}\varepsilon \tag{10-45}$$

and

$$-\widetilde{\Delta\mu^\circ} = \tilde{n}\mathscr{F}\varepsilon^\circ \tag{11-39}$$

If the cell reaction is written in general terms as

$$bB + cC \rightarrow hH + iI \tag{9-1}$$

and noting that 10-25 is equivalent to

$$\mu - \mu^\circ = RT \ln \frac{a_H{}^h a_I{}^i}{a_B{}^b a_C{}^c} \tag{11-40}$$

we can get

$$\varepsilon - \varepsilon^\circ = -\frac{RT}{\tilde{n}\mathscr{F}} \ln \frac{a_H{}^h a_I{}^i}{a_B{}^b a_C{}^c} \tag{11-41}$$

This is a form of what is usually called the Nernst equation. The a's can be replaced by products of activity coefficients and molalities (or, if appropriate, molarities or mole fractions) but in the cases in which the a's refer to strong electrolytes, we must use relations such as 11-10 and 11-26.

For cell 10-42 the cell reaction per faraday is

$$\tfrac{1}{2}H_2 + AgCl \rightarrow Ag + HCl(aq) \tag{10-43}$$

For this reaction 11-41 becomes

$$\varepsilon - \varepsilon^\circ = -\frac{RT}{\tilde{n}\mathscr{F}} \ln \frac{a_{HCl} a_{Ag}}{a_H{}^{1/2} a_{AgCl}} \tag{11-42}$$

For the cell as used in the experiments of Harned and Ehlers from which the data in Table 10-4 are taken we can take the activities of H_2, AgCl and Ag as unity. This and the fact that $\tilde{n} = 1$ reduces the equation to

$$\varepsilon - \varepsilon^\circ = \frac{-RT}{\mathscr{F}} \ln a_{HCl} \tag{11-43}$$

Putting in numerical values for R, T, and $\mathscr{F}$ and changing to common logarithms we get

$$\varepsilon - \varepsilon° = -0.05916 \log a_{HCl} \qquad (11\text{-}44)$$

Then using activity coefficients

$$\varepsilon - \varepsilon° = -0.05916 \log \gamma_\pm^2 m^2 \qquad (11\text{-}45)$$

$$\varepsilon - \varepsilon° = -0.11832 \log \gamma_\pm - 0.11832 \log m \qquad (11\text{-}46)$$

Our problem is now fairly well specified. In order to use the data in Table 10-4 to calculate the γ's for the various solutions we must first determine $\varepsilon°$ by extrapolating the data to such great dilution that $\gamma_\pm = 1$. At first thought this might seem impossible because then $\log m \to -\infty$. However ε also $\to \infty$ and so we rearrange the equation to

$$\varepsilon° = 0.11832 \log \gamma_\pm + 0.11832 \log m + \varepsilon \qquad (11\text{-}47)$$

The terms on the two sides of 11-47 do not become infinite as $m \to 0$. We now use for $\log \gamma_\pm$ the approximate expression 11-34 and we call the approximate value of $\varepsilon°$ so defined $\varepsilon°'$. $I = m$ in this case so

$$\varepsilon°' = -0.0592 \frac{\sqrt{m}}{1 + \sqrt{m}} + 0.0592\, bm + 0.1183 \log m + \varepsilon \qquad (11\text{-}48)$$

Log $\gamma_\pm$ approaches zero as $m \to 0$ so the value of $\varepsilon°'$ extrapolated to $m = 0$ will be the required value of $\varepsilon°$.

Recall the typical equation of a straight line

$$y = kx + c$$

Now if we rearrange 11-48 to 11-49

$$\varepsilon - 0.0592 \frac{\sqrt{m}}{1 + \sqrt{m}} + 0.1183 \log m = \varepsilon°' - 0.0592\, bm \qquad (11\text{-}49)$$

the left-hand side of this equation is a function of only the measurables ε and m. If we call the left side of this equation y and set $x = m$, a plot of y against x should give us a straight line, to the extent that $\varepsilon°'$ is a constant, that is, to the extent that 11-34 correctly gives $\gamma_\pm$. The slope of the line should be $-0.0592\, b$ and the value of y for $m = 0$ should be $\varepsilon°$.

PROBLEM 11-6   Carry out the described graphical procedure to find $\varepsilon°$ and b for cell 10-43 using the data in Table 10-4. Latimer[15] gives $\varepsilon° = 0.2222$ v for this cell.

∷

PROBLEM 11-7   By finding the point on the curve plotted for Problem 11-6 which corresponds to $m = 0.100$, estimate ε and then evaluate

[15] W. M. Latimer, *Oxidation Potentials*, Second edition, Prentice-Hall, 1952.

$\gamma_\pm$ for a 0.100 m aqueous HCl solution at 25 °C by the use of this ε and ε° from Problem 11-6 put into 11-46. Evaluate $\gamma_\pm$ for this concentration also by 11-34 (using $b = 0.2$) and compare the two results. Robinson and Stokes[16] give $\gamma_\pm = 0.796$.

As another example we can consider the cell

$$M_xHg \mid MCl_2(m) \mid AgCl, Ag \qquad (11\text{-}50)$$

Here M is Ca, Sr, or Ba and M_xHg is a 2-phase amalgam of the alkaline earth metal. The concentration of the aqueous salt solution is m. For the passage of 2 faradays the cell reaction is

$$M_{(in\ amalgam)} + 2AgCl \rightarrow MCl_2(m) + 2Ag \qquad (11\text{-}51)$$

PROBLEM 11-8   Show that for two different salt concentrations m_1 and m_2 the potentials of the cell are related by

$$\varepsilon_2 - \varepsilon_1 = \frac{-3RT}{2\mathscr{F}} \ln \frac{\gamma_{\pm_2} m_2}{\gamma_{\pm_1} m_1} \qquad (11\text{-}52)$$

::

PROBLEM 11-9   If ε^* is the potential of the cell when the activity of MCl_2 is 1, show that the potential of the cell when the molality of MCl_2 is m is given by the relation

$$\varepsilon - \varepsilon^* = -0.0886 \log \gamma_\pm - 0.0886 \log m - 0.0533 \qquad (11\text{-}53)$$

::

PROBLEM 11-10   By following a procedure similar to that described above show that

$$\varepsilon^* - 0.0886\, bm = z - 0.0886 \frac{\sqrt{3m}}{1 + \sqrt{3m}} \qquad (11\text{-}54)$$

where

$$z = \varepsilon + 0.0886 \log m + 0.0533 \qquad (11\text{-}55)$$

::

The data of Tippetts and Newton[17] for cell 11-50 with calcium amalgam and $CaCl_2$ solution measured at 25 °C are shown in the first two columns of Table 11-3. The rest of the table shows values calculated for some of the quantities shown in 11-54 (y in the last column is the entire right-hand side of 11-54). The data are plotted in Figure 11-1.

[16] R. A. Robinson and R. H. Stokes, *Electrolyte Solutions*, Butterworth Scientific Publications 1955.

[17] E. A. Tippetts and R. F. Newton, *J. Am. Chem. Soc.*, **56**, 1675 (1934).

Table 11-3

m	ε	$-0.0886 \log m$	z	$0.0886 \dfrac{\sqrt{3m}}{1 + \sqrt{3m}}$	y
0.0500	2.0453	0.1150	1.9836	0.0247	1.9598
0.0563	2.0418	0.1117	1.9843	0.0257	1.9586
0.0690	2.0348	0.0936	1.9853	0.0277	1.9576
0.1159	2.0183	0.0829	1.9887	0.0328	1.9559
0.1194	2.0175	0.0817	1.9889	0.0332	1.9559
0.1303	2.0147	0.0783	1.9895	0.0341	1.9556
0.1539	2.0094	0.0720	1.9905	0.0358	1.9549
0.2373	1.9953	0.0625	1.9861	0.0406	1.9455

PROBLEM 11-11   From Figure 11-1 estimate ε^* and b.

::

PROBLEM 11-12   By interpolation on Figure 11-1 estimate y for $m = 0.100$. From this value and your value for ε^* estimate ε for the cell for $m = 0.100$. Using ε^* and ε in 11-53 calculate a value for $\gamma_\pm$ in the 0.100 m solution.

::

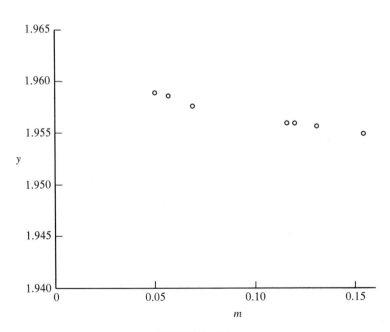

FIGURE 11-1

PROBLEM 11-13 Calculate $\gamma_\pm$ for the 0.100 m solution by 11-34. By comparing the answers to Problems 11-12 and 11-13, judge the validity of the statement previously attributed to Davies that 11-34 should be good up to $I = 0.1$ within about 2%. Harned and Owen[18] list two values, 0.492 and 0.495, for $\gamma_\pm$ for 0.100 m $CaCl_2$ at 25°.

::

PROBLEM 11-14 N.B.S. Circular 500 (reference previously cited) gives for the standard free energy of formation of AgCl $\Delta G_f^\circ = -26.244$ kcal mole^{-1}. It gives for the standard free energy of formation of aqueous $CaCl_2$ (hypothetical 1 m solution is standard state) $\Delta G_f^\circ = -194.88$ kcal mole^{-1}. Calculate ε° for the reaction

$$Ca + 2AgCl \rightarrow CaCl_2(aq) + 2Ag \qquad (11\text{-}56)$$

::

PROBLEM 11-15 Consider the double cell

$$Ca\,|CaCl_2(a = 1)|\,Ag,\,AgCl\,|CaCl_2(a = 1)|\,Ca_xHg \qquad (11\text{-}57)$$

Write the equation for the change in state corresponding to the passage of two faradays through this cell. Using the value of ε° calculated in Problem 11-14 and the value of ε^* estimated in Problem 11-11, calculate the activity of Ca in a 2-phase amalgam (Ca_xHg) at 25 °C.

HALF-CELL POTENTIALS

We have already discussed the reasons that make it impossible to determine on any generally accepted basis the activities or activity coefficients of individual ions. The same reasons make it impossible, without an arbitrary convention, to get experimental values for the ΔH's, ΔG's, ΔS's, etc., of formation of individual ions. It is very convenient, however, to adopt a convention which permits the tabulation of sets of consistent values of the thermodynamic properties of formation of individual ions. This convention is[19] that the standard properties of formation for reaction 11-58 are zero.

$$\tfrac{1}{2}H_2 \rightarrow H^+(aq) + e^- \qquad (11\text{-}58)$$

That is, for 11-58 it is taken that ΔH°, ΔG°, ΔS°, etc., are all zero. 11-58 can be called a half-reaction because electrons do not exist free in aqueous solutions—the electron produced in 11-58 must be taken up by another half-reaction such as 11-59

$$e^- + \tfrac{1}{2}Cl_2 \rightarrow Cl^-(aq) \qquad (11\text{-}59)$$

---

[18] H. S. Harned and B. B. Owen, *The Physical Chemistry of Electrolytic Solutions*, Reinhold Publishing Corp., 1950.

[19] The kind of treatment which follows is used chiefly in aqueous solutions although it is applicable to other ionizing but unionized solvents. It is not very useful in fused salt solutions.

PROBLEM 11-16   By adding 11-58 and 11-59 show that the standard free energy of formation of $Cl^-(aq)$ is just the standard free energy of formation of HCl(aq) (see 11-7).

::

Standard half-cell potentials can be assigned through the same convention plus the equation

$$\widetilde{\Delta\mu}^\circ = \Delta G^\circ = -\tilde{\eta}\mathscr{F}\varepsilon^\circ \tag{11-60}$$

When one uses the term "electrode potential of the chlorine–chloride electrode," the convention adopted by the International Union of Pure and Applied Chemistry is that the potential meant is the reduction potential. A reduction potential, in turn, means the potential of a cell in which the anode is a hydrogen–hydrogen ion electrode and the cathode is the electrode to which the reduction potential refers. However, inasmuch as various conventions have been used from time to time, one would be well advised just to say that he is referring to an oxidation potential or to a reduction potential, or preferably to write the equation such as 11-58 or 11-59 to which the half-cell potential refers.

PROBLEM 11-17   The standard free energy of formation of aqueous HCl is -31.372 kcal mole^{-1}. Calculate the standard reduction potential (electrode potential) of the Cl_2, Cl^- electrode, i.e., the standard potential of half-reaction 11-59.

::

PROBLEM 11-18   For the reaction

$$\tfrac{1}{2}H_2 + \tfrac{1}{2}Cl_2 \rightarrow HCl(aq) \tag{11-61}$$

$\Delta S^\circ = -28.8$ cal deg^{-1} mole^{-1}. What other data are needed to calculate S° for $Cl^-(aq)$? S° for $Cl^-(aq) = 13.5$ cal deg^{-1} mole^{-1}.

::

ADDITIONAL PROBLEMS

PROBLEM 11-19   The properties of formation of an aqueous solution of a strong electrolyte may be considered as the sum of the properties of formation of the ions into which the electrolyte dissociates. For example, ΔG_f° for HCl(aq) is the sum of the ΔG_f°'s for H^+ and Cl^- ions. Using Table 10-2 (see p. 235) and conventions discussed above, calculate ΔG_f° for Na^+.

::

PROBLEM 11-20 Use Table 10-2 and relevant principles to calculate standard reduction potentials for the electrodes:

a. $e^- + CuCl(c) \rightarrow Cu + Cl^-(aq)$

b. $2e^- + Cl_2(g) \rightarrow 2Cl^-(aq)$

c. $2e^- + 2H^+(aq) + HgO(c) \rightarrow Hg(l) + H_2O(l)$

::

PROBLEM 11-21 In the neutralization of a strong acid with a strong base, the acid ion and the base ion undergo no change. The neutralization reaction is

$$H^+(aq) + OH^-(aq) \rightarrow H_2O(l)$$

Using Table 10-2 calculate for the neutralization of a strong acid with a strong base: ΔH°_{298}, ΔG°_{298}, ΔS°_{298}.

::

PROBLEM 11-22 Using your estimated value for ε^* (see Problem 11-11) and assuming that $b = 0.2$ in Equation 11-34, plot y against m and compare the graph with the experimental result represented by Figure 11-1.

::

PROBLEM 11-23 Calculate the ionization constant of water at 25° and at 100 °C.
Answer: K at $100° = 0.95 \times 10^{-12}$

::

PROBLEM 11-24 Show that for cases in which the reference state is one of infinite dilution, the standard chemical potential can be defined by the equation

$$\mu^\circ = \lim (\mu - RT \ln y)$$

where $y = P, C, m$, or X, as appropriate.

::

PROBLEM 11-25 From the data given relative to the cells shown in Figure 10-2, and using $\gamma_\pm = 0.657$ for $1.022m$ NaCl, calculate ε° for $Na^+ + e^- \rightarrow Na$. Compare your answer with your answer to Problem 11-19.

12

Gibbs–Duhem Equations

The Gibbs–Duhem relation is the basic thermodynamic link between the properties of one component of a solution and the corresponding properties of the other components. Gibbs–Duhem equations can be written with respect to partial molal values of any property. For example, relations can be written among the partial molal volumes, the partial molal enthalpies, or the chemical potentials of the various components of a solution. The relations among chemical potentials are especially useful. It is these relations and their expressions in term of activities and of osmotic and activity coefficients with which this chapter is chiefly concerned.

DERIVATION OF GIBBS–DUHEM RELATIONS

Hildebrand and Scott[1] give a simple and rather unusual derivation of the Gibbs–Duhem relation. Consider two different ways of making an infinitesimal change in the composition of a solution of A and B. The first is to add dn_A moles of A. For this process the increment in the chemical potential of A would be given by

$$d\mu_A = \left(\frac{\partial \mu_A}{\partial n_A}\right) dn_A \tag{12-1}$$

Another process by which an identical change in the composition of the system can be brought about is first to (a) add an amount of the same

[1] J. H. Hildebrand and R. L. Scott, *The Solubility of Nonelectrolytes*, Third edition, Reinhold Publishing Corp., 1950. Several other ways of making the derivation are commonly given and can be found in books on physical chemistry and chemical thermodynamics. See also Problems 13-49 through 13-53.

solution containing dn_A moles of A and dn_B moles of B and then (*b*) to distill out or otherwise remove the dn_B moles of B which were added together with the dn_A moles of A. Inasmuch as the composition of the solution is not changed by the first part of this process (*a*), the total change in μ_A will correspond to that produced by the second part of this process (*b*) which is given by

$$d\mu_A = -\left(\frac{\partial \mu_A}{\partial n_B}\right) dn_B \tag{12-2}$$

Then we write

$$\mu_A = \frac{\partial G}{\partial n_A} \tag{12-3}$$

and

$$\frac{\partial \mu_A}{\partial n_B} = \frac{\partial^2 G}{\partial n_A \, \partial n_B} \tag{12-4}$$

But G is a property (dG is an exact differential) so we can apply the reciprocity relation (see 6-16)

$$\frac{\partial^2 G}{\partial n_A \, \partial n_B} = \frac{\partial \mu_B}{\partial n_A} \tag{12-5}$$

Thus

$$\frac{\partial \mu_B}{\partial n_A} = \frac{\partial \mu_A}{\partial n_B} \tag{12-6}$$

Combining 12-1, 12-2, and 12-6 gives

$$\left(\frac{\partial \mu_A}{\partial n_A}\right) dn_A + \left(\frac{\partial \mu_B}{\partial n_A}\right) dn_B = 0 \tag{12-7}$$

But, because dn_A and dn_B were the increments of the respective components resulting from the addition of an infinitesimal amount of the solution in which the mole fractions were X_A and X_B, we can write

$$\frac{dn_A}{dn_B} = \frac{X_A}{X_B} \tag{12-8}$$

This, when combined with 12-7 gives

$$X_A \, d\mu_A + X_B \, d\mu_B = 0 \tag{12-9}$$

12-9 is a form of the Gibbs–Duhem relation applied to the chemical potential. Although it is most useful for 2-component systems it can be generalized to

$$X_A \, d\mu_A + X_B \, d\mu_B + X_C \, d\mu_C + \cdots = 0 \tag{12-10}$$

The relation can be applied to any partial molal property. As applied to partial molal volumes, for example,

$$X_A \, d\bar{V}_A + X_B \, d\bar{V}_B + \cdots = 0 \tag{12-11}$$

ACTIVITY RELATIONS IN 2-COMPONENT SOLUTIONS

Our basic definition of the activity ratio is 10-19

$$\mu_B'' - \mu_B' = RT \ln \frac{a_B''}{a_B'} \tag{10-19}$$

Differentiation gives for the infinitesimal change

$$d\mu = RT \, d \ln a \tag{12-12}$$

Considering a 2-component system, a combination of 12-9 and 12-12 gives

$$X_A \, d \ln a_A + X_B \, d \ln a_B = 0 \tag{12-13}$$

One useful relation that follows easily from 12-13 has to do with vapor pressures. In differential form 10-34 is

$$RT \, d \ln a = \bar{V} \, dP \tag{12-14}$$

If P_B is taken to be the vapor pressure of component B, and to the extent that it is a satisfactory approximation to treat the vapors as ideal gases, 12-14 becomes

$$d \ln a_B = d \ln P_B \quad \begin{bmatrix} \text{If vapor} \\ \text{is ideal} \\ \text{gas} \end{bmatrix} \tag{12-15}$$

PROBLEM 12-1   Suppose that A is the solvent and B the solute in a 2-component solution and that A obeys Raoult's law

$$P_A = P_A{}^0 X_A \tag{10-12}$$

By combining 12-13, 12-15, and 10-12 show that the following equations can be obtained.

$$d \ln P_B = \frac{-X_A}{1 - X_A} \, d \ln (P_A{}^0 X_A) \tag{12-16}$$

$$d \ln P_B = \frac{1}{1 - X_A} \, d(1 - X_A) \tag{12-17}$$

By carrying out an indefinite integration of 12-17 after substituting X_B for $1 - X_A$ show that

$$\ln \frac{P_B}{X_B} = \text{constant} \tag{12-18}$$

thus proving that if, over a certain range of compositions, Raoult's law holds for the solvent and the vapors are ideal gases, Henry's law must hold for the solute over this range of compositions.

PROBLEM 12-2   Prove the converse of Problem 12-1, namely that if the solute obeys Henry's law the solvent must obey Raoult's, if the vapors are ideal.

::

12-13 can be rearranged and an integration indicated as follows

$$\ln \frac{a''_A}{a'_A} = -\int_{'}^{''} \frac{X_B}{X_A} \, d \ln a_B \tag{12-19}$$

The integration is to be taken with the values of the variables X_A, X_B, and a_B going from what they are in the state of the system indicated by $'$ to what they are in the state indicated by $''$. A sometimes more useful form of the relationship indicated by 12-19 can be written in terms of the γ's defined by

$$a = \gamma X \tag{10-27}$$

PROBLEM 12-3   By substituting 10-27 into 12-19 show that

$$\ln \frac{\gamma''_A}{\gamma'_A} = -\int_{'}^{''} \frac{X_B}{1 - X_B} \, d \ln \gamma_B \tag{12-20}$$

::

It is evident that 12-19 and 12-20 can be written equally correctly in terms of common logarithms. Thus

$$\log \frac{a''_A}{a_A} = -\int_{'}^{''} \frac{X_B}{1 - X_B} \, d \log X_B \tag{12-21}$$

$$\log \frac{\gamma''_A}{\gamma'_A} = -\int_{'}^{''} \frac{X_B}{1 - X_B} \, d \log \gamma_B \tag{12-22}$$

PROBLEM 12-4   An expression equivalent to 12-19 is

$$\ln \frac{a''_A}{a'_A} = -\int_{'}^{''} \frac{n_B}{n_A} \, d \ln a_B \tag{12-23}$$

Show that

$$\ln \frac{a''_A}{a'_A} = \frac{-M_A}{1000} \int_{'}^{''} m \, d \ln a_B \tag{12-24}$$

where m is the molality of B in the solution and M_A is the molecular weight of the solvent.

::

PROBLEM 12-5   Fortunately, we usually do not get into trouble by treating as a strong electrolyte one which may be rather far from complete dissociation at some of the concentrations of interest. If, for example, we treat H_2SO_4 in this way (even though HSO_4^- may be largely undissociated at concentrations of interest) we would get an expression for its activity analogous to the one for $BaCl_2$ (see 11-32), namely

$$a_{H_2SO_4} = 4\gamma_\pm^3 m^3 \tag{12-25}$$

By putting 12-25 into 12-24 show that for an aqueous sulfuric acid solution

$$\log \frac{a''_{H_2O}}{a'_{H_2O}} = -0.0540 \int_{m'}^{m''} m \, d \log \gamma_\pm - 0.0234(m'' - m') \tag{12-26}$$

∷

PROBLEM 12-6 Table 12-1 gives values for $\gamma_\pm$ for aqueous sulfuric acid solutions of various concentrations obtained by Harned and

Table 12-1 Mean Activity Coefficient of Sulfuric Acid in Aqueous Solutions at 25 °C*

m	$\gamma_\pm$	m	$\gamma_\pm$	m	$\gamma_\pm$
0.0005	0.885	0.02	0.453	2	0.124
0.0007	0.857	0.03	0.401	5	0.212
0.001	0.830	0.05	0.340	10	0.553
0.002	0.757	0.07	0.301	15	1.093
0.003	0.709	0.1	0.265	17.5	1.471
0.005	0.639	0.2	0.209		
0.007	0.591	0.5	0.154		
0.01	0.544	1.0	0.130		

* H. S. Harned and W. J. Hamer, *J. Am. Chem. Soc.*, **57**, 27 (1935). Reprinted by permission of the American Chemical Society.

Hamer[2] from emf measurements. By using these data and graphically integrating 12-26, estimate the activity of water in a $1m$ H_2SO_4 solution at 25 °C.

∷

PROBLEM 12-7 In Problems 10-14 and 10-15 two extreme interpretations of the nature of a moderately concentrated aqueous solution of H_2SO_4 were considered. One was to treat the solute as undissociated, the other was to treat it as completely dissociated into H^+ and $SO_4^=$ ions. A third and probably better approximation that might be used would be to treat the first ionization ($H_2SO_4 \rightarrow H^+ + HSO_4^-$) as complete and to ignore the second ($HSO_4^- \rightarrow H^+ + SO_4^=$). From the data of Scatchard, Hamer, and Wood (see Table 12-2) one can calculate that the activity of water in a $1m$ H_2SO_4 solution at 25 °C is 0.9620. Using this figure and calculating X_{H_2O} both on the basis (*a*) of complete dissociation of the H_2SO_4 and (*b*) of

[2] H. S. Harned and W. J. Hamer, *J. Am. Chem. Soc.*, **57**, 27 (1935)

dissociation only to H^+ and HSO_4^-, calculate the activity coefficient of water $(a = \gamma X)$ in a 1 m H_2SO_4 solution.

Answers: (a) 1.015, (b) 0.997

::

Although Problem 12-6 illustrates the use of a Gibbs–Duhem relation, it represents the hard way of getting the activity of water in a 1m H_2SO_4 solution; it can be determined directly and accurately by vapor pressure measurements. However, there are difficulties in applying the Gibbs–Duhem relation in the other way in order to use water vapor pressure measurements to evaluate activity coefficients for H_2SO_4 and other solutes. One of these difficulties depends upon the fact that the reference state for solutes is taken as the infinitely dilute solution. Thus, if one is doing a graphical integration of 12-19 or 12-20, one of the limits of integration will be where $X_B/X_A \rightarrow \infty$ (taking A in this case to refer to the solute). Another difficulty is that γ's for ionic solutes remain far from unity even at concentrations so low that the vapor pressure lowering of the solvent will be so small as to be difficult to measure to at least several significant figures. Later, we consider in some detail the use of freezing-point data, a procedure which can be thought of as an indirect method of making very precise vapor pressure measurements. At present we consider a slightly less indirect method, that of isopiestic equilibration.

OSMOTIC COEFFICIENTS

Before describing this method and giving some examples of the use of iso-piestic data, however, we define the osmotic coefficient.[3] Problem 12-7 has illustrated the fact that the activity coefficient of the solvent may, in some cases, be a clumsy variable, not only because of the ambiguity of mole fractions in cases of solutes which dissociate but also because the very small departures of these activity coefficients from unity may fail completely as indications of the departure of the solutions from ideal behavior. In intro-ducing their discussion of osmotic coefficients, Robinson and Stokes[4] use an example similar to that of Problem 12-7. They point out that at 25° in a 2m KCl solution the activity coefficient of the water is 1.004 whereas $\gamma_\pm$ for KCl is 0.614.

[3] Proposed by N. Bjerrum, *Z. Electrochem.*, **24**, 259 (1907).
[4] R. A. Robinson and R. H. Stokes, *Electrolyte Solutions*, Butterworth Scientific Publications, London, 1955. This book is an excellent reference not only for theory, data, and bibliographies relevant to its subject matter, but also for valuably in-formative descriptions of some experimental methods capable of giving precise data. With particular reference to our present considerations it should be pointed out that these authors are among the pioneers in the refinement of methods of isopiestic and "bithermal" equilibration.

We shall only suggest the rationale for the definition of the practical osmotic coefficient as given in 12-34. For a Raoult's law solution over which the solvent vapor may be considered an ideal gas[5] the activity of the solvent will be

$$a_{\text{Solvent}} = \frac{n_{\text{Solvent}}}{n_{\text{Solvent}} + n_{\text{Solute}}} \quad [\text{Ideal}] \tag{12-28}$$

Long division gives

$$a_{\text{Solvent}} = 1 - \frac{n_{\text{Solute}}}{n_{\text{Solvent}} + n_{\text{Solute}}} \tag{12-29}$$

For dilute solutions this leads to the approximate relations

$$a_{\text{Solvent}} \approx 1 - \frac{n_{\text{Solute}}}{n_{\text{Solvent}}} \tag{12-30}$$

$$\ln a_{\text{Solvent}} \approx - \frac{n_{\text{Solute}}}{n_{\text{Solvent}}} \tag{12-31}$$

and

$$\ln a_{\text{Solvent}} \approx \frac{-mM_{\text{Solvent}}}{1000} \tag{12-32}$$

where M_{Solvent} is the molecular weight of the solvent and m is the molality of the solution. For an m molal ideal dilute solution of a solute which dissociates completely into $m\nu$ moles of dissociation products ($\nu = x + y$ in 11-25) 12-32 becomes

$$\ln a_{\text{Solvent}} \approx \frac{-m\nu M_{\text{Solvent}}}{1000} \tag{12-33}$$

The practical osmotic coefficient ϕ, a convenient measure of the departure of the solvent from ideal behavior in nonideal dilute solutions, is then defined by

$$\ln a_{\text{Solvent}} = -\phi \frac{m\nu M_{\text{Solvent}}}{1000} \tag{12-34}$$

In fact ϕ as defined by 12-34 can be used advantageously with reference to solutions which are not very dilute and in which the solute does not approach complete dissociation. In such cases it is not ϕ itself but the difference between ϕ and ϕ_{Ideal} which indicates the degree of departure of the solvent from ideal behavior.

::

[5] Another way of describing the qualifications would be to say that the solutions obeys Rault's law expressed in terms of fugacity

$$f_{\text{Solvent}} = f^{\circ}_{\text{Solvent}} X_{\text{Solvent}} \tag{12-27}$$

PROBLEM 12-8   Using 12-28 and 12-34 show that for a $1m$ aqueous solution of a 1–2 electrolyte (such as Na_2SO_4) $\phi =_{Ideal} 0.98$ [6]

∷

It should be evident that a Gibbs–Duhem relation must exist between the osmotic coefficient, in terms of which the chemical potential (activity) of the solvent can be expressed, and the activity coefficient of the solute, in terms of which its chemical potential (activity) can be expressed.

∷

PROBLEM 12-9   To get convenient forms of this relationship start by differentiating 12-24 to get

$$d \ln a_{Solvent} = \frac{-mM_{Solvent}}{1000} d \ln a_{Solute} \qquad (12\text{-}36)$$

Then using an equation like 11-29 for a_{Solute} and 12-34 for $\ln a_{Solvent}$ show that

$$d \ln \gamma_{\pm} = d\phi + (\phi - 1)d \ln m \qquad (12\text{-}37)$$

By examining your solution to this problem observe that 12-37 is valid also for a nondissociated solute, in which case $\gamma_{\pm}$ is replaced in 12-37 simply by γ. In the integrations of 12-37 which follow, therefore, we use γ to represent either $\gamma_{\pm}$ for an electrolyte or γ for a nonelectrolyte.[7]

∷

PROBLEM 12-10   Show that a correct expression for the activity coefficient of the solute in terms of its molality and the osmotic coefficient is

$$\ln \gamma = \phi - 1 + \int_{m=0}^{m=m'} \frac{\phi - 1}{m} dm \qquad (12\text{-}38)$$

∷

[6] It may seem strange to define a reference function ϕ which does not remain unity for ideal systems at all concentrations. There is nothing incorrect, however, about doing so, and such a definition simplifies dealing with concentrations expressed in molality. The rational osmotic coefficient $\phi_{Rational}$ is defined by

$$\ln a_{Solvent} = \phi_{Rational} \ln X_{Solvent} \qquad (12\text{-}35)$$

$\phi_{Rational}$ will have a value of unity for solutions of any concentration to which 12-28 applies.

[7] Although there are many examples of nonelectrolytes which dissociate into uncharged dissociation products, these are almost invariably treated on the basis that concentrations both of undissociated and of dissociated species can be measured and that appropriate activity coefficients can be defined and are subject to thermodynamic measurement for each species. That is, no mean activity coefficient like $\gamma_{\pm}$ is used in these cases.

PROBLEM 12-11 Starting with 12-24 it follows immediately that

$$\ln a_{\text{Solvent}} = \frac{-M_A}{1000} \int_{m=0}^{m=m'} m \, d \ln a_{\text{Solute}} \tag{12-39}$$

This leads to

$$\ln a_{\text{Solvent}} = \frac{-M_A}{1000} \int_0^{m'} \nu m \, d \ln \gamma_m \tag{12-40}$$

Show that by putting the defining equation for ϕ into 12-40 one can derive that

$$\phi = 1 + \frac{1}{m'} \int_{m=0}^{m'} m \, d \ln \gamma \tag{12-41}$$

::

PROBLEM 12-12 If one were to evaluate the integral in 12-38 graphically one would plot $(\phi - 1)/m$ against m for values of m from 0 to m'. By using 12-41 together with the simplest Debye–Hückel limiting law

$$- \ln \gamma = \text{constant} \times \sqrt{m}$$

show that $(\phi - 1)/m$ does not remain finite but goes to $-\infty$ as $m \to 0$.

::

PROBLEM 12-13 By noting that $dm^{1/2} = (1/2m^{1/2}) \, dm$ show that 12-38 can be rearranged to

$$\ln \gamma = \phi - 1 + 2 \int_0^{m'} \frac{\phi - 1}{\sqrt{m}} \, d\sqrt{m} \tag{12-42}$$

Show that for a Debye–Hückel solute $(\phi - 1)/\sqrt{m}$ remains finite as $m \to 0$. Derive an expression for $\phi - 1$ in the Debye–Hückel limiting law region.

::

The principle of isopiestic equilibration is very simple. But as with many kinds of experiments which test and use simple theory, the refinements which allow the method to be used with great precision are tributes to the skill of those who have worked in the field. Even so, the useful range of the method does not extend to concentrations in aqueous solution much less than $0.1m$.

In this method, one dish contains a solution of one nonvolatile solute and another dish contains a solution of a different nonvolatile solute both in the same volatile solvent; the dishes are at kept the same temperature and in communication through an enclosed vapor space. In due course the solvent will evaporate and condense so as to make the vapor pressure of the solvent in the two dishes equal. In the equilibrium situation, therefore, the activities of the solvent in the two dishes must be equal.

Table 12-2 Osmotic Coefficients and Logarithms of Activity Coefficients at 25 °C*

	NaCl		KCl		H₂SO₄		Sucrose	
m	ϕ	$1 + \log \gamma$	ϕ	$1 + \log \gamma$	ϕ	$1 + \log \gamma$	ϕ	$1 + \log \gamma$
0.1	0.9342	0.8928	0.9264	0.8865	0.6784	0.4239	1.0073	1.0063
0.2	0.9255	0.8678	0.9131	0.8566	0.6675	0.3202	1.0151	1.0130
0.3	0.9224	0.8531	0.9063	0.8377	0.6685	0.2620	1.0234	1.0198
0.4	0.9217	0.8430	0.9023	0.8240	0.6723	0.2225	1.0319	1.0270
0.5	0.9224	0.8358	0.9000	0.8134	0.6773	0.1931	1.0407	1.0343
0.6	0.9242	0.8304	0.8987	0.8048	0.6824	0.1700	1.0497	1.0418
0.7	0.9266	0.8265	0.8980	0.7978	0.6895	0.1521	1.0590	1.0494
0.8	0.9295	0.8236	0.8980	0.7918	0.6980	0.1380	1.0684	1.0572
0.9	0.9329	0.8214	0.8982	0.7867	0.7075	0.1269	1.0781	1.0652
1.0	0.9363	0.8199	0.8985	0.7822	0.7176	0.1184	1.0878	1.0732
1.2	0.9434	0.8135	0.8996	0.7751	0.7396	0.1061	1.1076	1.0895
1.4	0.9509	0.8179	0.9008	0.7685	0.7634	0.0997	1.1280	1.1062
1.6	0.9589	0.8188	0.9024	0.7635	0.7888	0.0977	1.1484	1.1231
1.8	0.9681	0.8209	0.9048	0.7596	0.8154	0.0992	1.1686	1.1400
2.0	0.9786	0.8243	0.9081	0.7568	0.8431	0.1034	1.1884	1.1567
2.5	1.0096	0.8366	0.9194	0.7532	0.9152	0.1229	1.2359	1.1978
3.0	1.0421	0.8530	0.9330	0.7532	0.9922	0.1525	1.2817	1.2382
3.5	1.0783	0.8727	0.9478	0.7557	1.0740	0.1902	1.3262	1.2778
4.0	1.1168	0.8951	0.9635	0.7599	1.1606	0.2346	1.3691	1.3166
4.5	1.1578	0.9199	0.9799	0.7655			1.4100	1.3542
5.0	1.2000	0.9464	(0.9900)	(0.7695)ᵃ			1.4477	1.3902
5.5	1.2423	0.9742					1.4820	1.4244
6.0	1.2861	1.0029					1.5070	1.4540
	(1.2987)	(1.0114)ᵃ					(1.5090)	(1.4568)ᵃ

ᵃ Saturated solution.
* From G. Scatchard, W. J. Hamer, and S. E. Wood *J. Am. Chem. Soc.*, **60**, 3069 (1938). Used by permission of the American Chemical Society.

Scatchard, Hamer, and Wood[8] present the data reproduced in Table 12-2. The values of ϕ and $\log \gamma$ were calculated at the rounded molalities shown from the results of their own isopiestic experiments together with other kinds of data obtained by themselves and by others. The usefulness of these supplementary data is shown very clearly by the fact that for the electrolytes even at the lowest concentrations at which it was found practical to use the isopiestic method, the activity coefficients have not nearly approached one. So, for example, from the data reproduced here in Table 12-1, by the methods we used to get the answer to Problem 11-12 they calculated values for the γ's of H₂SO₄ that could be extrapolated to $m \to 0$. From these through 12-41 the corresponding ϕ's could be calculated.

Alternatively, they used values for the γ's of NaCl solutions obtained from freezing point and heat of dilution data. The values so calculated could then be compared with those calculated for the H₂SO₄ solutions by methods we develop shortly. Similar independent calculations were made on the KCl solutions. Thus, the data in Table 11-2 represent a critical evaluation and intercomparison of all the relevant experimental results available to the authors.

[8] George Scatchard, W. J. Hamer, and S. E. Wood, *J. Am. Chem. Soc.*, **60**, 3069 (1938).

The relation between the ϕ's for two isotonic solutions is easy to derive. From the defining equation for ϕ and the fact that $\ln a_{\text{Solvent}}$ is the same by the definition of isotonic solutions of two solutes B and C, we can write

$$\nu_B m_B \phi_B = \nu_C m_C \phi_C \tag{12-43}$$

and if we define the isopiestic ratio R by

$$R = \frac{\nu_B m_B}{\nu_C m_C} \tag{12-44}$$

we get

$$\phi_C = R\phi_B \tag{12-45}$$

PROBLEM 12-14   The experimental data on which the values in Table 12-2 are based include the fact that a $2.5022m$ solution of NaCl is isotonic with a $1.9998m$ solution of H_2SO_4. From this fact and the appropriate data from Table 12-2 determine whether H_2SO_4 is being treated as a 1–1 strong electrolyte $(H_2SO_4 \rightarrow H^+ + HSO_4^-)$ or as a 1–2 type $(H_2SO_4 \rightarrow 2H^+ + SO_4^=)$. Calculate R for the pair of isotonic solutions cited.

::

PROBLEM 12-15   From 12-42 one can conclude that if one plots ϕ against $\nu m \phi$ for a series of solutes, vertical lines will cut the several curves at points corresponding to values of ϕ in isotonic solutions. Make two such graphs from the data in Table 12-2, one covering values of m from 0–1 and the other for values of $m > 1$.

::

PROBLEM 12-16   For an ideal aqueous solution use 12-28 to calculate a_{Solvent} from for several values of νm. From these through 12-34 calculate the corresponding values of $\nu m \phi$ and of ϕ. Add to the graphs drawn in Problem 12-13 a curve representing the ideal solution.

::

ACTIVITY COEFFICIENT RELATIONS IN ISOTONIC SOLUTIONS

From 12-36 we can write for a pair of isotonic solutions of the solutes B and C

$$\frac{1000}{M_{\text{Solvent}}} d \ln a_{\text{Solvent}} = m_B \, d \ln a_B = m_C \, d \ln a_C \tag{12-46}$$

The second equation in 12-46 leads to

$$\nu_B m_B \, d \ln (m_B \gamma_B) = \nu_C m_C \, d \ln (m_C \gamma_C) \tag{12-47}$$

Using the definition of R in 12-43 we get

$$d \ln \gamma_C + d \ln m_C = R \, d \ln \gamma_B + R \, d \ln m_B \qquad (12\text{-}48)$$

This can be rearranged to

$$d \ln \gamma_C + d \ln m_C = d \ln \gamma_B + d \ln m_B + (R - 1) \, d \ln (\gamma_B m_B) \quad (12\text{-}49)$$

Indicating the integration gives

$$\ln \gamma_C = \ln \gamma_B + \int_0^{m_B'} d \ln \frac{m_B}{m_C} + \int_0^{m_B'} (R - 1) \, d \ln (\gamma_B m_B) \quad (12\text{-}50)$$

The first integral can be evaluated by noting that as $m_B \to 0$, for a pair of isotonic solutions, $R = 1$, i.e.,

$$\nu_B m_B = \nu_C m_C \quad \begin{bmatrix} \text{For infinitely} \\ \text{dilute isotonic} \\ \text{solutions} \end{bmatrix} \qquad (12\text{-}51)$$

Hence, the first integral in 12-50 becomes

$$\int_0^{m_B'} d \ln \frac{m_B}{m_C} = \left[\ln \frac{m_B}{m_C} \right]_{\frac{m_B}{m_C} = \frac{\nu_C}{\nu_B}}^{\frac{m_B}{m_C} = \frac{m_B'}{m_C'}} = \ln \frac{\nu_B m_B'}{\nu_C m_C'} = \ln R' \qquad (12\text{-}52)$$

Hence we get from 12-50 and 12-52

$$\ln \gamma_C = \ln \gamma_B + \ln R' + \int_0^{m_B'} (R - 1) \, d \ln (\gamma_B m_B) \qquad (12\text{-}53)$$

(In the foregoing equations R and m_B are variables whereas R' and m_B' are used to represent the values of R and m_B in the isotonic solutions having the activity coefficients γ_C and γ_B.)

Unfortunately again, and perhaps somewhat drearily, we must concede that there is no simple and accurate way even with 12-53 to get activity coefficients from isopiestic data alone, or even from isopiestic data involving one solute whose activity coefficients have been accurately determined by some means to very low concentrations. The difficulty is that the integration in 12-53 must be done all the way to $m_B = 0$ and values of R will not be known accurately much below $m = 0.1$. An equation like 12-53 can, however, be used to compare data in the isopiestic range and to calculate activity coefficients at various concentrations if a value is known for one concentration and if the necessary isopiestic data are available with respect to a standard solute whose activity coefficients are known over the whole range of interest.

PROBLEM 12-17   Let γ'_C and γ''_C be the activity coefficients of C in two solutions which are isotonic with solutions of B in which the activity coefficients are γ'_B and γ''_B. From 12-53 show that it follows that

$$\ln \frac{\gamma''_C}{\gamma'_C} = \ln \frac{\gamma''_B}{\gamma'_B} + \ln \frac{R''}{R'} + \int_{m_B'}^{m_B''} (R - 1)\, d \ln (\gamma_B m_B) \qquad (12\text{-}54)$$

::

PROBLEM 12-18   It should be possible to get ϕ for a $4m$ sucrose solution easily to several significant figures from vapor pressure measurements. Assume that the value quoted in Table 12-2 has been so obtained. Use the graph plotted in Problem 12-15 to find the molality of a solution of H_2SO_4 which is isotonic with $4m$ sucrose. Using H_2SO_4 as the standard solute calculate ϕ for a $1m$ sucrose solution using 12-45 and the graph of Problem 12-15. Compare the answer with Table 12-2.

::

PROBLEM 12-19   In this problem we let C refer to H_2SO_4 and B to KCl. In 12-54 let γ'_C refer to the activity coefficient of the solute in $0.1m$ H_2SO_4 and γ'_B to the corresponding isotonic solution (see graph of Problem 12-15) of KCl. We assume that these γ's are correctly given by Table 12-2, interpolating for γ'_B. Now use 12-54 to calculate by graphical integration a value for the activity coefficient of H_2SO_4 in its $2.0m$ solution assuming that KCl can be used as a standard, i.e., that the values of γ_B which would be for KCl solutions isotonic with various H_2SO_4 solutions are correctly given in Table 12-2. Compare your answer with the value of γ for $2m$ H_2SO_4 given in Table 12-1.

::

PROBLEM 12-20   Use 12-38 and the osmotic coefficient data in Table 12-2 to calculate γ for sucrose in a $5m$ solution. Compare your answer with the value of $\log \gamma$ given in Table 12-2.

::

PROBLEM 12-21   If the standard state of sucrose were taken as the pure crystal instead of as a hypothetical $1m$ solution based on the infinitely dilute reference state, values of γ could be calculated from the osmotic coefficients without extrapolation. Taking the concentration of a solution of sucrose saturated at $25°$ as 6.053 molal, use the data in Table 12-2 to calculate, with the use of an appropriately modified form of 12-38, the value of γ based on the pure crystal standard state for (*a*) a $1m$ and (*b*) a $0.1m$ sucrose solution. Compare your answers with the values listed in the table by noting that $\dfrac{a''}{a'}$ must be the same for a pair of solutions regardless of the choice of standard state.

::

PROBLEM 12-22   For an aqueous solution of a single electrolyte 11-40 can be written

$$\ln \gamma = A\left(\frac{\sqrt{Bm}}{1 + \sqrt{Bm}} - 0.2\, Bm\right) \tag{12-55}$$

where A and B vary with the valence type of the electrolyte. By setting $Bm = x^2$ we get

$$\ln \gamma = A\left(\frac{x}{1 + x} - 0.2x^2\right) \tag{12-56}$$

From this

$$d \ln \gamma = A\left(\frac{1}{(1 + x)^2} - 0.4x\right) dx \tag{12-57}$$

We had as an expression for ϕ

$$\phi = 1 + \frac{1}{m'} \int_0^{m'} m\, d \ln \gamma \tag{12-41}$$

Show that if the concentration is low enough so that 12-55 satisfactorily gives $\ln \gamma$, then

$$\phi - 1 = \frac{A}{B}\left(1 + x - 2 \ln (1 + x) - \frac{1}{1 + x} - 0.1x^4\right) \tag{12-58}$$

::

PROBLEM 12-23   Using values of A and B that can be obtained from previously stated relations show that 12-58 would predict for $0.1m$ aqueous solutions of 1–1, 1–2, and 2–1 type electrolytes at 25°: $\phi_{1-1} = 0.930$, $\phi_{1-2} = \phi_{2-1} = 0.854$. This result agrees well with the values shown in Table 12-2 for KCl and NaCl but not for H_2SO_4. Some other values of ϕ for $0.1m$ aqueous solutions as quoted by Robinson and Stokes are: HCl 0.943, NaOH 0.925, $NaClO_4$ 0.930, NH_4NO_3 0.911, $MgCl_2$ 0.861, $CaCl_2$ 0.854, $Ca(ClO_4)_2$ 0.883, $Cu(NO_3)_2$ 0.847, K_2SO_4 0.779, $(NH_4)_2SO_4$ 0.767, K_2CrO_4 0.805, $K_2Cr_2O_7$ 0.868.

::

The isopiestic method is just another example of chemical potentiometry. In it the vapor space serves as a conductor which permits one component to achieve equality of chemical potential in two systems to be compared, without letting the other components pass.

13

Solutions. Phase and Selective-Barrier Equilibria

THE GIBBS PHASE RULE

For a system which is at a uniform temperature and pressure and is composed of a number of phases, the criterion that the phases are in equilibrium with each other is that the chemical potential of each component is not different in one phase from what it is in any other. As has been stated before, an equally definitive and somewhat more mechanistic criterion is that the vapor pressure of each component should be invariant from phase to phase. Thus $P'_A = P''_A = P'''_A = \cdots$ and $P'_B = P''_B = P'''_B = \cdots$ etc. Here the letters A, B, ... denote the various components and the primes denote the various phases. One can conclude that there are $c(p - 1)$ independent equations of this type, c meaning the number of components and p the number of phases. Each such independent equation lessens by one the number of parameters of the system which are independently variable. The number of independently variable parameters is often called the number of degrees of freedom, f.

    Next we ask how many independent parameters there would be if there were no requirements representing phase equilibria. The usual set of parameters is the temperature, the pressure, and the numbers specifying the composition of each phase. For a 2-component system one number (35% A, for example) will specify the composition of any particular phase. For a 3-component system it will take two numbers for each phase, etc. Thus, counting T and P, the total number of independent parameters if there were no phase-equilibria would be $p(c - 1) + 2$. Now if we subtract the number of phase equilibrium conditions, $c(p - 1)$, from the number of parameters, $p(c - 1) + 2$, which would be independently variable if there were no phase equilibria, we get the Gibbs phase rule:

$$f = c - p + 2 \tag{13-1}$$

Among the most interesting cases subject to the phase rule are those for which $f = 0$, the nonvariant systems. Of these, the simplest is the 1-component system. It is clear from 13-1 that a 3-phase 1-component system is nonvariant. This means that the T and P of such a system are unique properties of it and that neither can be arbitrarily set or changed without one of the phases disappearing. There is only one value of T and one of P at which, for example, ordinary ice, liquid water, and water vapor can coexist in the 1-component H_2O system. The temperature of this triple point is now the reference point of the Kelvin temperature scale and is defined as 273.1600 °K.[1]

An interesting system to consider with respect to the phase rule is one used as a standard of electrical potential. The saturated Weston standard cell is described by 13-2.[2]

$$HgCd_y + Hg_xCd\left(\begin{array}{c}\text{2-phase}\\\text{amalgam}\end{array}\right) |CdSO_4 \cdot \tfrac{8}{3}H_2O(\text{Crystals}| \, CdSO_4(\text{Satd soln})$$

$$|Hg_2SO_4(\text{Crystals})| \, Hg \tag{13-2}$$

As in other electrochemical systems we have already considered, absolute equilibrium (absence of any change or tendency to change) does not exist in the Weston cell. In this cell the solution cannot be saturated at all places with Hg_2SO_4 because in the presence of metallic cadmium the concentration of Hg_2^{++} will be reduced to a vanishingly small value by the reaction

$$Hg_2^{++} + Cd \rightarrow Cd^{++} + 2Hg \tag{13-3}$$

But because of the low solubility of Hg_2SO_4 in a saturated $CdSO_4$ solution, the rate of diffusion will be very small. Experimental evidence is convincing that thermodynamic treatment of many electrochemical cells gives extremely reproducible results and ones consistent with those obtained by other methods, in spite of the fact that there are always some irreversible processes going on in the cells. Perhaps there is no more remarkable example of such a cell than the Weston cell which under the best controlled conditions gives potentials reproducible to a few parts in a million, potentials which are not sensitive to the length of the diffusion path between electrodes.

[1] This is higher by almost exactly 0.01° than the temperature of the ice point (defined as the freezing point of air-saturated water under 1 atm air pressure). 25 °C is 25 °K above the ice point, i.e., is 298.15 °K.

[2] A description of the construction and characteristics of standard cells is given by Walter J. Hamer, "Standard Cells—Their Construction, Maintenance and Characteristics," National Bureau of Standards Monograph 84, 1965.

PROBLEM 13-1 Describe quantitatively the change in state accompanying the passage of 2 faradays through cell 13-2 (left electrode taken as anode). Consider the saturated $CdSO_4$ solution to be z molal and the solubility of Hg_2SO_4 to be very small.

::

PROBLEM 13-2 By counting the number of components and number of phases in the Weston cell one finds that $c = 5$ and $p = 6$. How does it happen that the cell can be put into a thermostat at the arbitrary temperature of 25 °C and under the arbitrary pressure of 1 atm without having any of the phases disappear? Equation 13-1 would give $f' = -1$ under these conditions, f' being the degrees of freedom remaining after two have been used up by arbitrarily specifying T and P.

::

PROBLEM 13-3 An answer that might be given to Problem 13-2 would be that we have here an electrical variable which was not considered present when 13-1 was derived. So for this case we should perhaps write $f = c - p + 3$. But for the cell to be stable, i.e., for the 6-phase system to persist at 25° and 1 atm we do not have to connect any wires to it. One might ask, then, in what respect this system differs from other examples of phase-equilibrium systems; in other words, why is the electrical variable not always relevant and why is it extraordinary to have to write the phase rule $f = c - p + 3$?

::

PROBLEM 13-4 What would be the final state of a Weston cell with respect to number and kind of phases present after the processes connected with the slow[3] diffusion of the Hg_2SO_4 to the Cd electrode had gone to completion?

::

THE CLAPEYRON EQUATION

One of the most useful phase-equilibrium relations is the Clapeyron equation. Consider a 2-phase, 1-component system at equilibrium. The phase rule

---

[3] We observed in deriving the phase rule that removal of a constraint (to the interaction between a pair of phases) lessened by one the number of degrees of freedom. Howard Reiss (*Methods of Thermodynamics*, Blaisdell Publishing Company, 1965) puts it the other way and points out that every constraint added to a system adds a degree of freedom—semantically, at least, certainly a paradox. Because of the unusual emphasis Reiss puts on the concept of constraints in developing the results of classical thermodynamics, one should find his book very interesting and thought-provoking.

One is tempted to pursue philosophically the idea that adding a constraint may add a degree of freedom. We possibly increase the temptation but resist it at the same time by pointing out only that certain constraints placed upon the interactions among individuals in a society may increase the freedom with which one can move about with impunity.

gives $f = 1$ for this case. This means that if we arbitrarily specify either T or P, the other must assume a value characteristic of the system if the two phases are to continue to coexist. For the transfer of substance from one phase to the other under equilibrium conditions, $\Delta G = 0$. This must be true under whatever set of T and P the two phases are in equilibrium. Suppose that the 2-phase system is initially at equilibrium at T and P. Both pressure and temperature are now changed infinitesimally. The change in ΔG will be (see 5-24 and 5-25)

$$d \Delta G = \Delta V \, dP - \Delta S \, dT \qquad (13\text{-}4)$$

But if the phases are to remain in equilibrium at $T + dT$ and $P + dP$, ΔG must remain 0 and thus $d \Delta G$ must be zero. Hence the relation between equilibrium temperature and pressure for such a system is

$$\frac{dP}{dT} = \frac{\Delta S}{\Delta V} \qquad \begin{bmatrix} \text{A form of} \\ \text{the Clapeyron} \\ \text{equation} \end{bmatrix} \qquad (13\text{-}5)$$

ΔS is the entropy increment and ΔV the volume increment corresponding to a transfer of some specified amount (a mole or a gram, for example—but the ΔS and ΔV must refer to the same amounts) of substance from one phase to the other.

　　Both ΔS and ΔV will in general vary with changes in the equilibrium conditions (T and P). But for a particular T and P 13-5 can be rewritten

$$\frac{dP}{dT} = \frac{\Delta H}{T \, \Delta V} \qquad \begin{bmatrix} \text{The} \\ \text{Clapeyron} \\ \text{equation} \end{bmatrix} \qquad (13\text{-}6)$$

Here ΔH is the enthalpy and T the temperature of the phase transition, the ΔH of vaporization and the boiling point, for example.

　　To use either 13-5 or 13-6 to calculate the equilibrium relations corresponding to finite changes in T and P an integration must be performed. The simplest approximation, valid for small changes in T and most useful for equilibria not involving a vapor phase, is to treat the right-hand side of 13-5 or 13-6 as a constant.[4] This approximation gives

$$\Delta P = \frac{\Delta H}{T \, \Delta V} \Delta T \qquad (13\text{-}7)$$

PROBLEM 13-5　Use 13-7 to estimate how much of the 0.01 degree difference between the normal freezing point of water and its triple point is dependent upon the pressure difference (the rest would be attributed to the effect of the dissolved air). Take the heat of fusion of ice as 80 cal g^{-1},

---

[4] Not much is to be gained by treating T as a variable under conditions under which ΔV and ΔH can be satisfactorily treated as constants.

the density of ice as 0.92 g cm^{-3} and the vapor pressure of water at the triple point as 4.579 torr.

Answer: About 0.0075° can be attributed to the pressure difference.

∷

The answer to Problem 13-5 indicates that for condensed-phase equilibria the variation in equilibrium pressure is very large compared to the corresponding variation in equilibrium temperature. Put the other way, we recognize a fact that we have already encountered, namely, that very large pressure changes are required to have much effect on condensed-phase equilibria.

The simplest approximations under which the Clapeyron equation can be satisfactorily integrated for solid-vapor or liquid-vapor equilibria are: that the vapor is an ideal gas, that ΔH is constant over the temperature range of interest, and that the volume of the liquid or solid phase can be neglected in comparison to the volume of the vapor.

PROBLEM 13-6   Show that when these approximations are put into 13-6 and the integration is performed one gets the alternative equations

$$\ln P = \frac{-\Delta H}{RT} + \text{constant} \tag{13-8}$$

$$\ln \frac{P_2}{P_1} = \frac{\Delta H}{R}\left(\frac{T_2 - T_1}{T_1 T_2}\right) \tag{13-9}$$

It should be borne in mind that in 13-8 and 13-9 the ΔH is the molal value because to get these equations the ideal gas law is written $PV = RT$. Equation 13-9 is often called the Clausius–Clapeyron equation.

∷

Equation 13-8 shows that to the extent that the approximations used in deriving it fit the facts, a plot of log P against $1/T$ should be a straight line the slope of which is $\Delta H/2.303R$. Actually, in treating experimental systems one gets a straighter line than might be expected. The explanation is that the nonconstancy of ΔH and the nonideality of the vapor partially compensate each other.

If density data are available it is possible to use a plot of $\ln P$ against $1/T$ to get values of ΔH of vaporization or sublimation without approximations. The slope of such a plot is

$$\text{Slope} = \frac{d \ln P}{d\left(\frac{1}{T}\right)} = -\frac{T^2}{P}\frac{dP}{dT} \tag{13-10}$$

When 13-6 is combined with 13-10 we get

$$\text{Slope} = -\frac{T\,\Delta H}{P\,\Delta V} \tag{13-11}$$

13-6 is a rigorous thermodynamic equation and 13-10 is a mathematical fact, therefore 13-11 is exact.

PROBLEM 13-7   The data in Table 13-1 are from Dorsey's comprehensive and critical tabulations in "The Properties of Ordinary Water-Substance."[5] Check the internal consistency of Table 13-1 by calculating

Table 13-1   Properties of the H_2O Liquid-Vapor Equilibrium System.*

t	P	dP/dT	ΔH_v	δ
0	0.0060273	0.4373	2500.00	103
10	0.012102	0.8103	2476.47	139
20	0.023042	1.4207	2452.98	109
30	0.041831	2.4015	2429.30	78
40	0.072748	3.8796	2405.54	60
50	0.121698	6.041	2381.58	48
60	0.196560	9.100	2357.37	51
70	0.307520	13.305	2332.83	36
80	0.467396	18.932	2307.87	32.6
90	0.691923	26.288	2281.42	29.2
100	1.00000	35.699	2256.66	26.4
150	4.69746	125.96	2113.73	17.8
200	15.3472	321.13	1940.08	13.22
250	39.2557	662.9	1714.77	10.64
300	84.7931	1194.8	1404.09	9.142
350	163.205	2002.0	892.86	8.58
374.15[a]	218.39	2630.7[b]	0	10.4

[a] Critical temperature   [b] At 374.0°

Units: t–°C; P–atm; dP/dT–milliatm deg^{-1}; ΔH_v–joules g^{-1}; the specific volume, cm^3 g^{-1}, is $\dfrac{82.05(t + 273.15)}{18.02P} - \delta$.

* These data are from N. E. Dorsey, "Properties of Ordinary Water-Substance," ACS Monograph Series #81, Reinhold Publishing Corporation, New York, 1940.

ΔH_v at 0° and 100 °C using 13-6 and the values of T, dP/dT and ΔV that can be obtained from columns 1, 2, 3, and 5 of the table. Verify that very little error is introduced by ignoring the volume of the liquid compared to that of the vapor at these temperatures.

::

PROBLEM 13-8   Make another check on the internal consistency of Table 13-1 by calculating the slope of a ln P vs $1/T$ plot both by 13-10 and 13-11. Make the calculations for two temperatures such as 0° and 150 °C.

::

[5] N. E. Dorsey, "The Properties of Ordinary Water-Substance," ACS Monograph Series #81, Reinhold Publishing Corporation, New York, 1940.

In many cases of solid-liquid equilibria one or more of the solid phases will be nearly pure substances even though the components are mutually soluble in the liquid phase. For example, NaCl has no appreciable solubility in ice and it will be very nearly pure crystalline water which is in equilibrium with NaCl solutions at their freezing points. Or if these solutions are saturated with NaCl one will have pure $H_2O(c)$ and pure $NaCl(c)$ in equilibrium with the saturated solution at its eutectic freezing point.

PROBLEM 13-9   Apply the phase rule to such a system and show that at the eutectic freezing point $f = 1$. The eutectic temperature is often thought of as a fixed temperature. Explain.

::

In cases of the kind described above, solutions have freezing points lower than those of the pure components. The Clausius–Clapeyron equation can be applied to the derivation of an approximate relation between the freezing point lowering and the concentration of the solution. The derivation will be for the freezing-point lowering of a component (A) which obeys Raoult's law:

$$P_A = P_A^\circ X_A \quad \text{[Raoult's law]} \tag{10-12}$$

Inasmuch as we need to identify various vapor pressures as well as two different temperatures, we specify the following notation.

T^* is the freezing point of pure A

T' is the freezing point of A in a solution in which its mole fraction $= X_A$

$P_{T_L*}^\circ$ is the vapor pressure of pure liquid A at T^*

$P_{T_S*}^\circ$ is the vapor pressure of pure solid A at T^*

$P_{T_L'}^\circ$ is the vapor pressure of pure liquid A at T'

$P_{T_S'}^\circ$ is the vapor pressure of pure solid A at T'

$P_{T_L'}$ is the vapor pressure of the solution at T'

PROBLEM 13-10   Verify the following relations applicable to the case under consideration:

$$\ln \frac{P_{T_L*}^\circ}{P_{T_L'}^\circ} = \frac{\Delta H_v}{R}\left(\frac{T^* - T'}{T'T^*}\right) \tag{13-12}$$

$$\ln \frac{P_{T_S*}}{P_{T_S'}^\circ} = \frac{\Delta H_s}{R}\left(\frac{T^* - T'}{T'T^*}\right) \tag{13-13}$$

$$P_{T_S'}^\circ = P_{T_L'} = P_{T_L'}^\circ X_A \tag{13-14}$$

$$P_{T_S*}^\circ = P_{T_L*}^\circ \tag{13-15}$$

$$-\ln X_A = \frac{\Delta H_s - \Delta H_v}{R}\left(\frac{T^* - T'}{T'T^*}\right) = \frac{\Delta H_{fs}}{R}\left(\frac{T^* - T'}{T'T^*}\right) \tag{13-16}$$

Here ΔH_v, ΔH_s, and ΔH_{fs} are the enthalpies of vaporization of liquid A, of sublimation of solid A and of fusion of solid A.

∷

To recapitulate the limitations on the validity of 13-16: it was required that no solid solutions be formed, that Raoult's law hold for A and that ΔH_s and ΔH_v be constants in the range T^* to T'. For some cases of mixtures of very similar substances, 13-16 might be a good approximation for all components over wide ranges of concentration. To illustrate this possibility we make use of a problem given by Moore.[6]

PROBLEM 13-11   In a 3-component system in accordance with the phase rule it is possible to have, at an arbitrarily chosen pressure, four phases in equilibrium with each other, but only at one particular temperature which is called the ternary eutectic temperature. One possible case would be that three of the phases are the pure crystalline forms of the three components and the fourth is a liquid solution containing all three components. This situation can be viewed as one in which the composition of the solution is such that the freezing points of each of the three components have been lowered to the same temperature.

The melting points and heats of fusion of o, p, and m dinitro-benzenes are[7]: 116.9 °C, 173.5 °C, 89.8 °C; and 3905, 3345, 4280 cal mole^{-1}. By setting up a set of simultaneous equations, estimate the ternary eutectic temperature and the composition of the ternary eutectic solution in this system.

Answer: $T = 46$ °C; $X_o = 0.33$; $X_p = 0.23$; $X_m = 0.44$

∷

A much commoner case to which 13-16 can be applied with good approximation is that of the dilute solution. It has been stated that in general Raoult's law as expressed by 10-12 becomes an increasingly good approximation as $X_A \to 1$. Thus 13-16 can be viewed as a limiting law for the freezing-point lowering of the solvent in dilute solutions. If it is to be used in this way 13-16 can be simplified by making use of other relations which hold more and more exactly as $X_A \to 1$.

PROBLEM 13-12   For a 2-component system (or for any number of components if we denote the total concentration of all solutes by X_B) $X_A = 1 - X_B$. By putting this relation into 13-16 show that a valid limiting law for dilute solutions is

$$X_B = \frac{\Delta H_{fs}}{R} \frac{\theta'}{T^{*2}} \tag{13-17}$$

in which $\theta' = T^* - T'$, the freezing-point lowering.

∷

[6] Walter J. Moore, *Physical Chemistry*, Third Edition, Prentice-Hall, Inc., 1962.

[7] Donald H. Andrews, *J. Phys. Chem.*, **29**, 882, 1041 (1925).

PROBLEM 13-13 By considering the definition of molality show that a limiting-law expression for the freezing-point lowering in dilute solutions is

$$\theta' = \frac{M_A R T_*^2}{1000 \, \Delta H_{fs}} \, m = \lambda m \tag{13-18}$$

Here we have replaced T^* by T_* for notational convenience. M_A is the molecular weight of the solvent. The second equation in 13-18 may be considered the definition of λ, the molal freezing-point lowering constant.

::

It can be seen that to the extent that 13-18 is a good approximation, the value of λ will be a constant for a given solvent and the value of θ' will be directly proportional[8] to m but not dependent upon the properties of the solute, provided that m can be evaluated to represent the total concentration of all solutes present, i.e., provided that association or dissociation processes are taken into account. Other phenomena the magnitudes of which, also in dilute solution, are proportional to m are the vapor pressure lowering, the boiling-point raising and the osmotic pressure. These four phenomena are sometimes called the colligative properties of solutions.

PROBLEM 13-14 By a derivation analogous to those in Problems 13-10, 13-12, and 13-13 show that the raising of the boiling point of a solvent by a nonvolatile solute, θ_b, is given on the basis of the same kinds of assumptions and approximations as those underlying 13-18 by

$$\theta_b = \frac{M_A R T_*^2}{1000 \, \Delta H_v} \, m = K_b m \tag{13-18A}$$

where T_* is the boiling point of the solvent and K_b is the molal boiling-point raising constant. Observe that the vapor pressure of the solvent over the pure solvent at its boiling point is equal to the vapor pressure of the solvent over the solution at its boiling point.

::

PARTIAL MOLAL PROPERTIES

Before investigating the quantitative relationships of freezing-point lowering and related phenomena to the activities of solvent and solute we develop somewhat further the relationships among partial molal and related properties. We have already written

$$dG = V \, dP - S \, dT + \left(\frac{\partial G}{\partial n_1}\right)_{T,P,n_2\cdots} dn_1 + \left(\frac{\partial G}{\partial n_2}\right)_{T,P,n_1,n_3,\ldots} dn_2 + \cdots \tag{9-20}$$

[8] Both T_* and ΔH_{fs} will, in general, be functions of the pressure. The above statement assumes that the determination of θ' as a function of m is being done at a constant pressure.

The partial derivatives in 9-20 are by definition the partial molal free energies of the respective components.

$$\bar{G}_1 \equiv \left(\frac{\partial G}{\partial n_1}\right)_{T,P,n_2\cdots} \qquad \bar{G}_2 \equiv \left(\frac{\partial G}{\partial n_2}\right)_{T,P,n_1,n_3,\ldots} \tag{13-19}$$

Moreover, the chemical potential was defined by (see 9-4)

$$\mu_1 = \bar{G}_1, \qquad \mu_2 = \bar{G}_2, \quad \text{etc.} \tag{13-20}$$

and although the definition was introduced when ideal gas systems were under consideration, it is generally valid, as, indeed, are other identifications of the μ's such as 9-22, 9-23, 9-24, 9-25, and 9-26.

Other partial molal properties are similarly defined; the term is restricted to partial derivatives with T and P constant.

$$\bar{V}_i = \left(\frac{\partial V}{\partial n_i}\right)_{P,T,n_1\cdots \slashed{n}i,nj\cdots}$$

$$\bar{H}_i = \left(\frac{\partial H}{\partial n_i}\right)_{P,T,n_1\cdots \slashed{n}i,nj\cdots} \tag{13-21}$$

$$\bar{S}_i = \left(\frac{\partial S}{\partial n_i}\right)_{P,T,n_1\cdots \slashed{n}i,nj\cdots}$$

$$\text{etc.}$$

We have already derived one important relation among partial molal properties, the Gibbs–Duhem relation. We derived it with respect to the partial molal free energies (chemical potentials) but the derivation is equally good for other partial molal properties and would lead to other equations such as 12-11 and, for example,

$$X_A \, d\bar{H}_A + X_B \, d\bar{H}_B + \cdots = 0 \tag{13-22}$$

Another basic equation can be derived from consideration of a particular process of making up a solution of specified composition. We illustrate the derivation with respect to volume, but the result is valid for any property, i.e., the V and the $\bar{V}$'s in 13-21 can correctly be replaced by H and $\bar{H}$'s, G and $\bar{G}$'s, etc.

In the process to be considered we start with a solution of volume V' containing n_1' moles of component 1, n_2' moles of component 2, etc. We now add infinitesimal amounts of the different components, amounts related to each other in exactly the way that the n''s are related to each other, i.e., $dn_1/n_1' = dn_2/n_2' = \cdots$. The result is to produce an infinitesimal additional amount of solution of exactly the same composition as the original. The corresponding increment in volume, by the definition of the partial molal volumes, $\bar{V}_1, \bar{V}_2, \ldots$ will be

$$dV = \bar{V}_1 \, dn_1 + \bar{V}_2 \, dn_2 + \cdots \tag{13-23}$$

Because the quantity of solution is being augmented without changing its composition, the $\bar{V}$'s will remain constant regardless of the quantities added by this process. Hence we can integrate 13-19 under this condition to get

$$V - V' = (n_1 - n_1')\bar{V}_1 + (n_2 - n_2')\bar{V}_2 + \cdots \qquad (13\text{-}24)$$

But 13-24 is free of any assumed relation between V and V', hence

$$V = n_1\bar{V}_1 + n_2\bar{V}_2 + \cdots \qquad (13\text{-}25)$$

Moreover, V and the $\bar{V}$'s are properties of the solution and not at all dependent upon the method by which the solution is prepared. Hence 13-25 which we inferred by considering a particular method of preparation must hold regardless of the history of the solution.

Another important princip!e that can be used in dealing with partial molal properties may or may not seem intuitive. Intuition is very useful in science but it should be used only as a challenge to verify the intuitively believed proposition. The p..nciple is that all of the thermodynamic relations that exist among the extensive properties of closed systems such as among E, H, G, C_P, etc., exist in the same form among the corresponding partial molal properties. We verify this principle with respect to two examples; other verifications can be made similarly.

We can, for example, start with the definition of G

$$G = H - TS \qquad (5\text{-}8)$$

By differentiating this with respect to n_1 holding T, P, $n_2 \cdots$ constant, we get

$$\bar{G}_1 = \bar{H}_1 - T\bar{S}_1 \qquad (13\text{-}26)$$

Evidently 5-8 and 13-26 are analogous.

Let us now write

$$\left(\frac{\partial G}{\partial T}\right)_P = -S \qquad (5\text{-}25)$$

$$\left(\frac{\partial \bar{G}_1}{\partial T}\right)_{P,n_2 n_3 \cdots} = \frac{\partial}{\partial T}\left(\frac{\partial G}{\partial n_1}\right)_{T,P,n_2 \cdots} \qquad (13\text{-}27)$$

or

$$\left(\frac{\partial \bar{G}_1}{\partial T}\right)_{P,n_2 \cdots} = \frac{\partial^2 G}{\partial T \, \partial n_1} \qquad (13\text{-}28)$$

By the reciprocity relation,

$$\left(\frac{\partial \bar{G}_1}{\partial T}\right)_{P,n_2 \cdots} = \frac{\partial}{\partial n_1}\left(\frac{\partial G}{\partial T}\right)_{P,n_1 n_2 \cdots} = \frac{\partial}{\partial n_1}(-S) \qquad (13\text{-}29)$$

$$\left(\frac{\partial \bar{G}_1}{\partial T}\right)_{P,n_2 \cdots} = -\bar{S}_1 \qquad (13\text{-}30)$$

which is analogous to 5-25.

To get an expression for $[\partial(\bar{G}_1/T)]/\partial T$ we write

$$\frac{\partial\left(\dfrac{\bar{G}_1}{T}\right)}{\partial T} = -\frac{1}{T^2}\bar{G}_1 + \frac{1}{T}\frac{\partial\bar{G}_1}{\partial T} \tag{13-31}$$

Combining 13-26, 13-30, and 13-31 gives

$$\frac{\partial\left(\dfrac{\bar{G}_1}{T}\right)}{\partial T} = -\frac{\bar{H}_1}{T^2} \tag{13-32}$$

We can use 13-32 to find out how activities and activity coefficients vary with temperature. From 10-19 and the definition that in the standard state the activity is one we get

$$\bar{G} - \bar{G}^\circ = RT \ln a \tag{13-33}$$

Putting this with 13-32 gives

$$\frac{R\,\partial \ln a}{\partial T} = -\frac{\bar{H} - \bar{H}^\circ}{T^2} \tag{13-34}$$

The difference between the value of a partial molal property of a substance in the standard state and the value in some other state is called a relative partial molal property.[9] Thus, the relative partial molal enthalpy $\bar{L}$ is defined as

$$\bar{L} \equiv \bar{H} - \bar{H}^\circ \tag{13-35}$$

and 13-34 can be rewritten

$$\frac{\partial \ln a}{\partial T} = -\frac{\bar{L}}{RT^2} \tag{13-36}$$

The partial derivatives in 13-34 and 13-36 imply constancy of the concentration parameter (X, m, or C). It should be evident, then, that the temperature dependence of the activity coefficient will also be given by equations like 13-36. For an undissociated substance for which we write

$$a = \gamma X \tag{10-27}$$

or

$$a = \gamma m \tag{10-29}$$

for example, the result is simply

$$\frac{\partial \ln \gamma}{\partial T} = -\frac{\bar{L}}{RT^2} \tag{13-37}$$

[9] We define such terms as they are encountered. However, for additional discussion of definitions and relations between partial molal, relative partial molal, apparent molal, and relative apparent molal properties see Frederick D. Rossini, *Chemical Thermodynamics*, John Wiley and Sons, Inc., 1950. See also Vivian Barfield Parker, the introduction to "Thermal Properties of Aqueous Uni-Univalent Electrolytes," National Standard Reference Data Series National Bureau of Standards 2, 1965.

PROBLEM 13-15   Show that for an electrolyte which dissociates into ν ions ($\nu = 3$ for $BaCl_2$, for example)

$$\frac{\partial \ln \gamma_\pm}{\partial T} = -\frac{\bar{L}}{\nu RT^2} \tag{13-38}$$

To get the relation between a's or γ's at two different temperatures 13-36, 13-37 or 13-38 must be integrated.

∷

PROBLEM 13-16   For a $1m$ aqueous KCl solution at 25° Pitzer and Brewer give $\bar{L}_{KCl} = -176$ cal mole^{-1}. Robinson and Stokes give for a $1m$ KCl solution at its freezing point $\gamma_\pm = 0.589$. The freezing point may be taken as $-3.25\,°C$. Making the approximation that $\bar{L}_{KCl}$ has a constant value between the freezing point and 25 °C, calculate $\gamma_\pm$ for KCl in this solution at 25 °C and compare with the value of 0.604 given by Robinson and Stokes.

∷

For a better approximation than that of Problem 13-16 we must express $\bar{L}$ as a function of T. We know that

$$\left(\frac{\partial H}{\partial T}\right)_P = C_P \tag{1-30}$$

Hence

$$\left(\frac{\partial \bar{L}}{\partial T}\right)_P = \bar{C}_P - \bar{C}_P^\circ \equiv \bar{J} \tag{13-39}$$

$\bar{J}$ is the symbol used for relative partial molal heat capacity. For a range of temperature over which $\bar{J}$ is approximately constant we would have

$$\bar{L} \approx l + \bar{J}T \tag{13-40}$$

Of course $\bar{J}$ can be treated as a function of T, the simplest nonconstant approximation would be a linear relation

$$\bar{J} \approx j + kT \tag{13-41}$$

In using 13-40 and 13-41 l, j, and k would be treated as functions of the concentration but not of the temperature.

APPARENT MOLAL PROPERTIES

A useful link between experimental data and calculated values of the partial molal and relative partial molal properties is the *apparent molal* property. The simplest illustration is the apparent[10] molal volume, Φ_V. Suppose that

---

[10] The symbol Φ used with V, etc. to indicate an apparent molal property should not be confused with ϕ which we have used for the osmotic coefficient.

a solution contains n_A moles of A, n_B moles of B, etc. Now suppose that we remove the n_B moles of B at constant T and P and leave the numbers of moles of the other components unchanged. The apparent molal volume of B in the original solution (1) would be defined in terms of the difference in volume between solution 1 and solution 2 as

$$\Phi_{V_B} = \frac{V_1 - V_2}{n_B} \quad \begin{bmatrix} \text{Number of moles of B in solution} \\ 1 = n_B, \text{ in solution } 2 = 0 \end{bmatrix} \quad (13\text{-}42)$$

It can be seen that as $n_B \to 0$

$$\Phi_{V_B} = \bar{V}_B \quad \text{[At infinite dilution of B]} \quad (13\text{-}43)$$

We now derive some relations between the apparent molal properties and the partial molal properties which are not restricted to the infinitely dilute solution. Although the definition of Φ_V and other apparent molal properties could be applied to any component of a solution, these quantities are usually measured and calculated with respect to solutes. When one expresses solution composition in molality or when one uses the infinitely dilute solute reference state, the distinction between solvent and solute becomes essential. As our sample of an apparent molal property we continue to use Φ_{V_B} which, in such cases, refers to a solute.

The total volume of the solution is given by 13-44 to be

$$V = n_A \bar{V}_A + n_B \bar{V}_B + n_C \bar{V}_C + \cdots \quad (13\text{-}44)$$

If we designate by $V^{\not{B}}$ the volume of a solution differing from the above only in that $n_B = 0$, we have

$$V^{\not{B}} = n_A \bar{V}_A^{\not{B}} + n_C \bar{V}_C^{\not{B}} + \cdots \quad (13\text{-}45)$$

and

$$\Phi_{V_B} = \frac{V - V^{\not{B}}}{n_B} \quad (13\text{-}46)$$

Hence

$$\Phi_{V_B} = \frac{V - n_A \bar{V}_A^{\not{B}} - n_C \bar{V}_C^{\not{B}} - \cdots}{n_B} \quad (13\text{-}47)$$

PROBLEM 13-17 By combining 13-44 and 13-47 show that

$$\Phi_{V_B} = \bar{V}_B + \frac{1}{n_B} [n_A(\bar{V}_A - \bar{V}_A^{\not{B}}) + n_C(\bar{V}_C - \bar{V}_C^{\not{B}}) + \cdots] \quad (13\text{-}48)$$

::

The apparent molal values of V, H, and C_P are rather directly accessible experimentally. This is because they can be calculated from

Table 13-2 Standard Enthalpies of Formation of H_2SO_4 in Aqueous Solutions at 25 °C, kcal mole^{-1}.*

n_{H_2O}	$-\Delta H_f$	n_{H_2O}	$-\Delta H_f$
0	194.548	500	212.833
1.0	217.32	1,000	213.275
1.5	201.193	5,000	214.390
2.0	203.128	10,000	215.060
3.0	206.241	50,000	216.545
5.0	208.288	100,000	216.855
10	210.451	500,000	217.187
20	211.500	1,000,000	217.246
100	212.150	∞	217.32

* From Wagman, Evans, Halow, Parker, Bailey, and Schumn, "Selected Values of Chemical Thermodynamic Properties," *N.B.S. Technical Note* 270-1 (1966).

volumetric or calorimetric measurements on the whole system. This is not true of free energies; vapor pressure, electrochemical, isopiestic, and other experiments of this kind lead primarily to values of the chemical potential of an individual component.

Table 13-2 gives molal values for ΔH_f° for H_2SO_4 in solutions of a number of different concentrations. A convention concerning such data should be pointed out. It is that the ΔH_f°'s tabulated represent the sum of ΔH_1 and ΔH_2 in Figure 13-1. This means that the enthalpy of formation of the solvent is not included in the value of the standard enthalpy of formation of the solute in solutions.

PROBLEM 13-18 From the data in Table 13-2 calculate the relative apparent molal heat content (relative apparent molal enthalpy), $\Phi_H - \Phi_{H^\circ}$ of H_2SO_4 in solution with 10 moles of water.

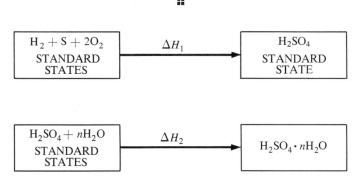

FIGURE 13-1

   PROBLEM 13-19   Craig and Vinal[11] give -0.233_2 kcal for the relative partial molal enthalpy (relative partial molal heat content) of water, $\bar{H}_{H_2O} - \bar{H}^\circ_{H_2O}$, in a solution containing ten moles of water per mole of H_2SO_4. Use this value together with the answer to Problem 13-18 to calculate the relative partial molal enthalpy of H_2SO_4 in this solution. This can be done by adapting 13-48 to enthalpies and to the 2-component system H_2O–H_2SO_4. To what do you attribute the discrepancy between your answer and the value given by Craig and Vinal which is $\bar{L}_{H_2SO_4} = 9.632$?

::

   We can rearrange 13-46 (and to emphasize the idea that these relations apply to any of the extensive properties we recast it in terms of H) to give

$$H = n_B \Phi_{H_B} + H^{\cancel{B}} \tag{13-49}$$

Taking the partial derivative with respect to n_B and recognizing that $H^{\cancel{B}}$ is not a function of n_B we get

$$\frac{\partial H}{\partial n_B} = \Phi_{H_B} + n_B \frac{\partial \Phi_{H_B}}{\partial n_B} \tag{13-50}$$

which for the constant T and P case is

$$\bar{H}_B = \Phi_{H_B} + n_B \frac{\partial \Phi_{H_B}}{\partial n_B} \tag{13-51}$$

   PROBLEM 13-20   Show that in a 2-component solution the relative partial molal property of component A is related to the apparent molal property of component B by an equation of the form

$$\bar{H}_A - \bar{H}^\circ_A = -\frac{n_2{}^2}{n_1} \frac{\partial \Phi_{H_B}}{\partial n_B} \tag{13-52}$$

(Refer to 13-48.) Note that the partial derivative of the apparent molal property which appears on the right-hand side of 13-52 can be replaced by the corresponding partial derivative of the relative apparent molal property, that is

$$\frac{\partial \Phi_{H_B}}{\partial n_B} = \frac{\partial \Phi_{L_B}}{\partial n_B} \tag{13-53}$$

::

   [11] D. Norman Craig and George W. Vinal, "Thermodynamic Properties of Sulfuric-Acid Solutions and Their Relation to the Electromotive Force and Heat of Reaction of the Lead Storage Battery," *Nat. Bur. Stds. J. of Research*, **24**, 475 (1940).

PROBLEM 13-21 By considering the relation between the numbers of moles of solvent and solute and the molality of a solution prove that for a 2-component solution 13-51 and 13-52 can be transformed to

$$\bar{H}_B = \Phi_{H_B} + m \frac{\partial \Phi_{H_B}}{\partial m} \tag{13-54}$$

$$\bar{H}_A - \bar{H}_A^\circ = -\frac{M_A m^2}{1000} \frac{\partial \Phi_{H_B}}{\partial m} \tag{13-55}$$

where M_A is the molecular weight of the solvent.

::

PROBLEM 13-22 It turns out that for strong electrolytes Φ_H and Φ_C (Φ_C is the apparent molal heat capacity at constant pressure) are comparatively linear functions of $\sqrt{m}$. In order to use 13-54 or 13-55 with such graphical data it is useful to have these equations transformed into ones in which the concentration variable is $\sqrt{m}$ instead of m. Show that such a

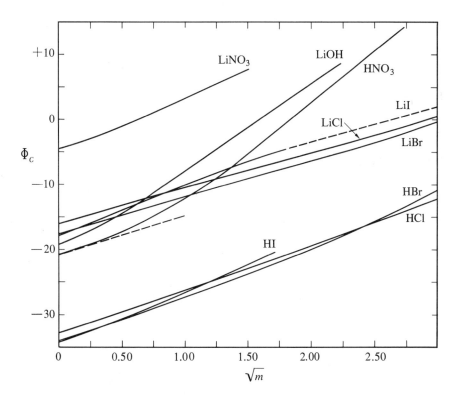

FIGURE 13-2 Apparent molal heat capacity (cal deg^{-1}) plotted against $\sqrt{m}$.

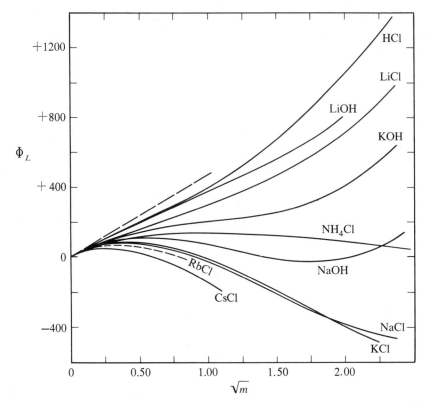

FIGURE 13-3.   Relative apparent molal enthalpy (calories) plotted against $\sqrt{m}$. Figures 13-2 and 13-3 are from "Thermal Properties of Aqueous Uni-Univalent Electrolytes" by Vivian Barfield Parker, National Standard Reference Data Series, National Bureau of Standards 2 (1964).

transformation leads to equations of the type

$$\bar{C}_B = \Phi_{C_B} + \tfrac{1}{2}m^{1/2}\frac{\partial \Phi_{C_B}}{\partial m^{1/2}} \tag{13-56}$$

$$\bar{L}_A = -\frac{M_A m^{3/2}}{2000}\frac{\partial \Phi_{H_B}}{\partial m^{1/2}} \tag{13-57}$$

where M_A is the molecular weight of the solvent and A and B always refer to solvent and solute respectively.

■■

PROBLEM 13-23   Figures 13-2 and 13-3 show the dependence of the apparent molal heat capacity and the relative apparent molal enthalpy on molality for aqueous solutions at 25 °C of a number of electrolytes. Use Figure 13-3 to estimate the relative partial molal enthalpy of NaCl in a 4m aqueous solution at 25°. Compare your answer with the value $\bar{L}_{NaCl} = -680$

cal mole^{-1} which can be estimated by interpolation of Parker's tabulated data.

::

PROBLEM 13-24 Use an equation like 13-44 together with information from Figure 13-2 to estimate how much the heat capacity of a very large quantity of 6*m* HNO_3 at 25° is increased by adding 1 mole of water.

::

To calculate partial molal properties from experimental data the graphical method of Roozeboom[12] is useful in some cases. Consider a 2-component solution of A and B. We designate by $\hat{V}$ that volume of such a solution which contains a total of 1 mole of A and B, i.e., for which $n_A + n_B = 1$. In this amount of solution, then $n_A = X_A$ and $n_B = X_B$ and from 13-25 we can write

$$\hat{V} = X_A \bar{V}_A + X_B \bar{V}_B \qquad (13\text{-}58)$$

Hence

$$\hat{V} = X_A \bar{V}_A + (1 - X_A)\bar{V}_B = (\bar{V}_A - \bar{V}_B)X_A + \bar{V}_B \qquad (13\text{-}59)$$

Suppose now that experimental values of $\hat{V}$ are plotted as ordinates against X_A. Remembering the typical equation of a straight line, we can see that if a straight line whose slope is $\bar{V}_A - \bar{V}_B$ is drawn through some point on the curve represented by 13-59, the straight line will be tangent to the curve at that point and its intercept with the ordinate at $X_A = 0$ will give the value of $\bar{V}_B$ corresponding to the point of tangency. Either on the basis of symmetry (the subscripts A and B can be interchanged in 13-59) or from the fact that the slope of the tangent is $\bar{V}_A - \bar{V}_B$ it can be concluded that the intercept on the ordinate where $X_B = 0$ will give the value of $\bar{V}_A$.

Figure 13-4 is a Roozeboom-type plot for aqueous sulfuric acid solutions of the molal constant-pressure heat capacity $\hat{C}$ against the mole fraction of water. It is based primarily on Biron's[13] determinations of the specific heats of sulfuric acid solutions of various concentrations.

PROBLEM 13-25 To see how the graph of Figure 13-4 was obtained from the experimental data, we reproduce a few lines of the data and make the corresponding calculations. In the tabulation below n_{H_2O} is

n_{H_2O}	% H_2SO_4	Specific heat	*d*
100	5.16	0.955	1.032
11	33.1	0.741	1.244
1	84.5	0.441	1.774
0.486	91.8	0.379	1.823

[12] H. W. B. Roozeboom, "Die Heterogenen Gleichgewichte," II-1, p. 288, Friedrich Vieweg and Sohn, Braunschweig, Germany, 1904.
[13] E. Biron, *Russkoe Fiz. Khim.* Obshehestvo, **31**, 201 (1899).

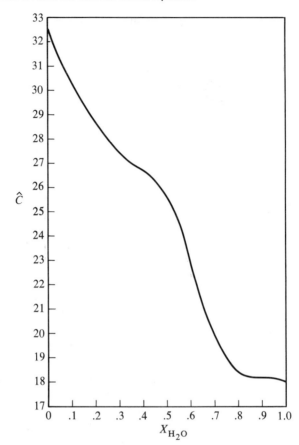

FIGURE 13-4   The molal heat capacity of constant pressure, $\hat{C}$, as a function of the mole fraction of H_2O in aqueous sulfuric acid solutions at 20°C. Units of $\hat{C}$ are cal deg^{-1}.

the number of moles of water per mole of H_2SO_4. The specific heats and densities d are in cal g^{-1} and g cm^{-3}. First calculate the weight of H_2O per mole of H_2SO_4. Then calculate the weight of 1 mole of solution. From the latter and the specific heat calculate $\hat{C}$. Calculate the corresponding values of X_{H_2O} and check your results against the graph.

::

PROBLEM 13-26   Use the tangent method described above together with Figure 13-4 to estimate values of $\bar{C}_{H_2SO_4}$ and $\bar{C}_{H_2O}$ in a 186m H_2SO_4 solution. Compare your results with the values calculated by Craig and Vinal (reference cited in Problem 13-19) which are $\bar{C}_{H_2SO_4} = 31.42$ and $\bar{C}_{H_2O} = 17.96$ cal. The Craig and Vinal values were calculated by a somewhat different method, namely one in which Φ_C values were first calculated and then equations of the form of 13-56 and 13-57 were used to calculate $\bar{C}_A$ and $\bar{C}_B$.

∷

PROBLEM 13-27   If c_P is the specific heat of an m molal aqueous solution of H_2SO_4 and c_P° is the specific heat of water show that the apparent molal heat capacity of H_2SO_4 in this solution, Φ_{C_B}, is given by

$$\Phi_{C_B} = \frac{1000}{m}(c_P - c_P^\circ) + M_B c_P \qquad (13\text{-}60)$$

where M_B is the molecular weight of H_2SO_4.

∷

PROBLEM 13-28   Use the Roozeboom method with the data in Table 13-2 to estimate $\overline{L}_{H_2O}$ and $\overline{L}_{H_2SO_4}$ in an aqueous solution of H_2SO_4 containing 10 moles of H_2O per mole of H_2SO_4. Compare your results with the values given and calculated in Problem 13-19.

∷

ACTIVITIES FROM FREEZING-POINT DATA

We are now prepared to study the relationships between the freezing-point lowering and the activities of solvent and solute. We consider a solution whose freezing point is $T_* - \theta'$, T_* being the freezing point of the pure solvent. Figure 13-5 is a change-in-state diagram which can help us to set up the basic relations. Referring to the numbered arrows in Figure 13-5 we note the following (always, here, with reference to the solvent):

$$\Delta \bar{G}_1 = 0 \qquad (13\text{-}61)$$

$$\Delta \bar{G}_{3'} = 0 \qquad (13\text{-}62)$$

$$\Delta \bar{G}_3 = \Delta \bar{G}_1 + \Delta \bar{G}_2 = \Delta \bar{G}_2 \qquad (13\text{-}63)$$

$$\Delta \bar{G}_2 = RT \ln a_A^* \qquad (13\text{-}64)$$

where a_A^* is the activity of the solvent in the solution at T_*.

$$\Delta \bar{H}_3 = \Delta H_{fs*} + \overline{L}_* \qquad (13\text{-}65)$$

where ΔH_{fs*} is the heat of fusion of the solvent at T_* and $\overline{L}_*$ is the relative partial molal enthalpy of the solvent in the solution at this temperature. At any other temperature we would have an analogous expression for $\Delta \bar{H}$ for the change in state $A(c) \rightarrow A(SOLN)$,[14] i.e.,

$$\Delta \bar{H} = \Delta H_{fs} + \overline{L} \qquad (13\text{-}66)$$

___

[14] It would not be possible isothermally at the pressure of the normal freezing point to carry out the change in state $A(c) \rightarrow A(l)$ below the normal freezing point to get ΔH_{fs} at $T_* - \theta$. One could take supercooled liquid to crystalline A at such a temperature. But in any case we can express ΔH_{fs} as a function of T in terms of ΔH_{fs*} and heat capacities which can be easily measured.

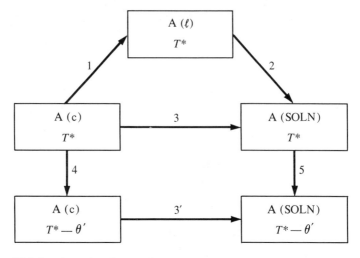

T^* is freezing point of pure solvent
$T^* - \theta'$ is freezing point of the solution
A (c) means crystalline solvent
A (ℓ) means pure liquid solvent
A (SOLN) means the solvent in solution

FIGURE 13-5   Freezing-point relations.

Inasmuch as $\Delta \bar{G}_{3'} = 0$ we can get an expression for ΔG_3 and hence $RT \ln a_A^*$ through the Gibbs–Helmholtz relation

$$\frac{\partial \left(\dfrac{\Delta \bar{G}}{T} \right)}{\partial T} = \frac{-\Delta \bar{H}}{T^2} \tag{13-32}$$

Let us express the temperature in terms of T_*, the freezing point of the pure solvent, and a variable θ related by the equation

$$T = T_* - \theta \tag{13-67}$$

Substituting into 13-32 gives

$$\frac{\partial \left(\dfrac{\Delta \bar{G}}{T} \right)}{\partial \theta} = \frac{\Delta \bar{H}}{(T_* - \theta)^2} \tag{13-68}$$

Combining 13-66 and 13-68 gives

$$\frac{\partial \left(\dfrac{\Delta \bar{G}}{T} \right)}{\partial \theta} = \frac{H_{fs} + \bar{L}}{(T_* - \theta)^2} \tag{13-69}$$

Indicating an integration (refer to Figure 13-5) we have

$$\frac{\Delta \bar{G}_3}{T_*} - \frac{\Delta \bar{G}_{3'}}{T_* - \theta'} = \int_{\theta=\theta'}^{\theta=0} \frac{\Delta H_{fs} + \bar{L}}{(T_* - \theta)^2} \, d\theta \qquad (13\text{-}70)$$

where $T_* - \theta'$ is the freezing point of the solution. At $T_* - \theta'$ the pure crystalline solvent is in equilibrium with the solution, hence $\Delta \bar{G}_{3'} = 0$. Also $\Delta \bar{G}_3 = \Delta \bar{G}_2$ (13-63). Taking these facts into account and combining 13-70 with 13-64 we get an expression for the activity of the solvent at its freezing point, namely

$$R \ln a_{\text{A}}^* = \int_{\theta=\theta'}^{\theta=0} \frac{\Delta H_{fs} + \bar{L}}{(T_* - \theta)^2} \, d\theta \qquad (13\text{-}71)$$

Before we can integrate 13-71 we have to consider the temperature dependence of ΔH_{fs} and $\bar{L}$. For ΔH_{fs} we can use 1-54 and write

$$\Delta H_{fs} = \Delta H_{fs*} + \int_{T*}^{T} \Delta C_P \, dT \qquad (13\text{-}72)$$

If we assume that ΔC_P varies linearly with the temperature, i.e.,

$$\Delta C_P = \Delta C_{P*} - k\theta \qquad (13\text{-}73)$$

and noting a fact we have already used (see 13-67)

$$d\theta = -dT \qquad (13\text{-}74)$$

we get, upon integrating 13-72

$$\Delta H_{fs} = \Delta H_{fs*} - \Delta C_{P*}\theta + \frac{k}{2} \theta^2 \qquad (13\text{-}75)$$

We get a similar expression for $\bar{L}$. We have already defined $\bar{J}$, the relative partial molal heat capacity,

$$\frac{d\bar{L}}{dT} = \bar{C}_P - \bar{C}_P^\circ \equiv \bar{J} \qquad (13\text{-}39)$$

Assuming a linear relation between $\bar{J}$ and the temperature over the range of interest we get, analogously to the ΔH_{fs} expressions,

$$\bar{J} = \bar{J}_* - b\theta \qquad (13\text{-}76)$$

and

$$\bar{L} = \bar{L}_* - \bar{J}_*\theta + \frac{b}{2} \theta^2 \qquad (13\text{-}77)$$

Comparing 13-75 and 13-77 with 13-71 shows that we can write

$$R \ln a_{\text{A}}^* = \int_{\theta=\theta'}^{\theta=0} \frac{A - B\theta + C\theta^2}{(T_* - \theta)^2} \, d\theta \qquad (13\text{-}78)$$

in which

$$\theta' = \text{freezing-point lowering}$$

$$A = \Delta H_{fs*} + \bar{L}_*$$

$$B = \Delta C_{P*} + \bar{J}_*$$

$$C = \frac{k + b}{2}$$

PROBLEM 13-29   Show that the result of integration of 13-78 is

$$R \ln a_A^* = -\frac{A\theta'}{T'T_*} + \frac{B\theta'}{T'} - B \ln \frac{T_*}{T'}$$

$$- C\theta' + 2CT_* \ln \frac{T_*}{T'} - C \frac{T_*}{T'} \theta' \qquad (13\text{-}79)$$

in which T' is the freezing point of the solution, i.e.,

$$T' = T_* - \theta'$$

::

PROBLEM 13-30   For water at 0 °C we can take $\Delta H_{fs*} = 1436$ cal mole^{-1} and $\Delta C_{P*} = 9.1$ cal deg^{-1} mole^{-1}. From the data tabulated by Craig and Vinal we can estimate for a 2.0m aqueous H_2SO_4 solution $\bar{L}_{25°} = 19$ cal mole^{-1} and $\bar{J}_{20°} = -0.2$ cal deg^{-1} mole^{-1}. The freezing point of a 2m H_2SO_4 solution is -10.14 °C.

Making the approximation that $\bar{L}$ and $\bar{J}$ are constant over the temperature range of interest and are equal to the above-quoted values, use 13-79 to estimate the activity of H_2O at 0 °C in a 2m H_2SO_4 solution. For comparison, the H_2O vapor pressure over pure water at 0 °C is 4.58 torr whereas from the tables of Blake and Greenewalt[15] one can estimate that the vapor pressure of water over 2m H_2SO_4 at 0 °C is 4.14 torr.

::

The solution to Problem 13-30 shows that the two B terms are nearly equal and of opposite sign—it may be difficult to get their difference with the desired precision. For reasons such as that, it is often useful to recast 13-79 into a rapidly converging series. To do this we start back with 13-78 and rewrite it

$$R \ln a_A^* = \int_{\theta=\theta'}^{\theta=0} \frac{A - B\theta + C\theta^2}{T_*^2 \left(1 - \dfrac{\theta}{T_*}\right)^2} d\theta \qquad (13\text{-}80)$$

---

[15] Frank C. Blake and C. H. Greenewalt, *International Critical Tables*, First Edition, Vol. 3, p. 302, Natl. Acad. Sci., 1928.

PROBLEM 13-31 Show that 13-80 can be written as the series

$$R \ln a_A^* = \frac{1}{T_*^2} \int_{\theta=\theta'}^{\theta=0} (A - B\theta + C\theta^2)\left(1 + \frac{2\theta}{T_*} + \frac{3\theta^2}{T_*^2} + \cdots\right) d\theta \quad (13\text{-}81)$$

and that upon integration and substitution of the limits one gets

$$R \ln a_A^* = -\frac{1}{T_*^2}\left\{A\theta' + \left(\frac{A}{T_*} - \frac{B}{2}\right)\theta'^2 + \left(\frac{A}{T_*^2} - \frac{2}{3}\frac{B}{T_*} + \frac{C}{3}\right)\theta'^3 + \cdots\right\}$$

$$(13\text{-}82)$$

::

PROBLEM 13-32 Solve Problem 13-30, but this time by the use of 13-82 instead of 13-79. Notice that the θ'^2 term of the series affects $R \ln a_A^*$ by only about 0.5% and the A and B parts of the θ'^3 term (we were not given the data for C) are about 4% of the θ'^2 contribution.

::

We have already seen that variations in the activity of the solvent are insensitive indicators of departures from ideality of dilute solutions. A related fact is that direct application of the Gibbs–Duhem equation to 13-80 or 13-82 is not very satisfactory. A better method is to work with the practical osmotic coefficient ϕ.[16] This we defined by the equation

$$\ln a_A = -\phi \frac{m\nu M_A}{1000} \quad (12\text{-}34)$$

It is obvious that we could put 12-34 directly into 13-82 (or 13-80) to get an expression for ϕ in terms of θ'. The dilute solution approximation comes out in simpler notation, however, if we introduce the molal freezing point lowering constant, λ, defined in 13-18.

PROBLEM 13-33 Bearing in mind the fact that $\bar{L}_*$ is the relative partial molal enthalpy of the solvent, show that a limiting law for dilute solution is

$$\phi = \frac{\theta'}{\lambda\nu m} \quad (13\text{-}83)$$

In the light of the results of Problem 13-32 we should expect 13-83 to be a good approximation for solutions up to molalities of at least a few tenths.

::

[16] Other functions can be used effectively. Lewis and Randall used the function j:

$$j = 1 - \frac{\theta}{\nu\lambda m}$$

Table 13-3   Freezing-Point Lowering in Aqueous KCl
Solutions.

m	θ'	m	θ'
0.01555	0.0558	0.08095	0.2806
0.03927	0.1381	0.08814	0.3048
0.06737	0.2343	0.1050	0.3615

PROBLEM 13-34   Table 13-3 gives the results of some very refined measurements made by P. M. G. Brown and J. E. Prue[17] of the freezing points of aqueous KCl solutions. Use 13-83 to calculate ϕ for the 0.105 m solution and compare with the value shown in Table 12-2.

⋮⋮

The principles to be used in treating concentrated solutions or in treating dilute solutions more elaborately than we did in Problem 13-34 are indicated by 13-80 and 13-82. We would not minimize the value of such treatments nor suggest that much ingenuity has not gone into their development. We shall not reproduce them here, however.

PROBLEM 13-35   To illustrate the calculation of solute activity coefficients from freezing-point data, we calculate $\gamma_{\pm}$ in a 0.105m KCl solution. First calculate values of ϕ for all the solutions of Table 13-3. Calculate also values for $\sqrt{m}$ and $(\phi - 1)/\sqrt{m}$. Then plot $(\phi - 1)/\sqrt{m}$ against $\sqrt{m}$ and from this plot and the calculated values of ϕ, using 12-42, calculate $\gamma_{\pm}$ for a 0.105m KCl solution. Compare with Table 12-2.[18]

⋮⋮

The answer to Problem 12-13, using the simplest form of the Debye–Hückel expression for $\gamma_{\pm}$, shows that as $m \to 0$, $\phi - 1$ should be proportional to $\sqrt{m}$. The graph of Problem 13-35 tends to bear this out, experimentally, inasmuch as the plot is fairly linear. However, the most reliable points on the graph are those farthest from the limit of $m \to 0$. For a 1–2 or 2–1 type electrolyte the problem of extrapolation would be even more serious. It is evident that for the most effective use of experimental data more sophisticated methods, such as those discussed by Monk, are in order.

SELECTIVE-BARRIER EQUILIBRIA

Without specifically classifying them as such we have already considered some important cases of selective-barrier equilibria. All of the electrochemical cells we have discussed belong in this category. In these cases certain processes

[17] P. M. G. Brown and J. E. Prue, *Proc. Roy. Soc.*, **232A**, 320 (1955).
[18] See C. B. Monk, *Electrolytic Dissociation*, Academic Press, 1961, for a detailed treatment of the KCl case including higher concentrations.

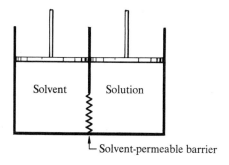

FIGURE 13-6   The osmotic pressure phenomenon.

have gone substantially to equilibrium whereas others that would tend to occur have been prevented, or more exactly, have been made negligibly slow by some kind of a barrier. In the Weston standard cell practical equilibrium is maintained with respect to saturation of the 2-phase Cd amalgam and with respect to saturation of the aqueous solution with $CdSO_4 \cdot \frac{8}{3}H_2O$. The process of migration of Hg_2SO_4 to the Cd electrode where it would be reduced by the Cd is, on the other hand, so slow as to be practically negligible in the thermodynamic treatment of the cell. The barrier in this case is the very small solubility of Hg_2SO_4 in the saturated $CdSO_4$ solution. Another important case of selective-barrier equilibrium is that of isopiestic equilibration. The barrier in that case is the vapor-phase path between the two solutions. This barrier selectively allows the volatile solvent to pass between the two dishes but makes the passage of the relatively nonvolatile solute negligibly slow.

The phenomenon of osmotic pressure is a classical case of selective-barrier equilibrium. The phenomenon can be described in terms of a selective barrier usually called a semipermeable membrane. This membrane has the property of being adequately permeable to the solvent but being practically impermeable to the solute(s). Figure 13-6 is a schematic diagram of an osmotic pressure apparatus. Pure solvent and a solution of some solute in that solvent are separated from each other by a barrier through which the solvent can pass but the solute cannot. The pressure on the two sides of the apparatus can be adjusted independently.

Let P^* be some reference pressure. Inasmuch as activities and activity coefficients are usually tabulated for 1 atm, P^* would usually be 1 atm. If P° is the pressure on the pure solvent and P the pressure on the solution, we have for the chemical potentials with respect to the reference pressure (taking the pure solvent at P^* as the standard state)

$$\mu^\circ_{A(\text{in pure solvent})} = \mu^*_A + \int_{P^*}^{P^\circ} V^\circ_A \, dP \qquad (13\text{-}84)[19]$$

[19] Note that in these equations we are using the superscript $^\circ$ to identify the pure solvent at any pressure but the combination $^\circ_*$ to identify the pure solvent at the pressure chosen for its standard state.

and

$$\mu_{A(\text{in solution})} = \mu_A^* + \int_{P^*}^{P} \bar{V}_A \, dP \tag{13-85}$$

in which V_A° is the molal volume of the pure solvent and $\bar{V}_A$ is the partial molal volume of the solvent in the solution; the subscript A identifies the solvent. Equilibrium with respect to the process we have assumed to be permitted, namely, that of passage of solvent through the barrier, will exist when there is no net tendency for the solvent to pass either way, and this condition will exist when $a_{A(\text{Solvent})} = a_{A(\text{Solution})}$. Thus at osmotic equilibrium $\mu_A = \mu_A^\circ$ and hence, by subtracting 13-85 from 13-84,

$$\mu_A^* - \mu_A^{\overset{\circ}{*}} = \int_{P^*}^{P^\circ} V_A^\circ \, dP - \int_{P^*}^{P} \bar{V}_A \, dP \tag{13-86}$$

But the left-hand member is $RT \ln a_A^*$, a_A^* being the activity of the solvent in the solution at P^*, so that 13-86 can be rewritten

$$RT \ln a_A^* = \int_{P^*}^{P^\circ} V_A^\circ \, dP - \int_{P^*}^{P} \bar{V}_A \, dP \tag{13-87}$$

The osmotic pressure Π is defined by

$$\Pi = P - P^\circ$$

when the system is at osmotic equilibrium and would be the pressure across the selective barrier (semipermeable membrane) at osmotic equilibrium. Note that experimentally either one of the integrals in 13-87 can be done away with. That is, either the pressure on the solution or the pressure on the pure solvent can be maintained at P^* and the osmotic equilibrium attained by adjusting the pressure on the other liquid. Keeping $P = P^*$ for the solution and adjusting the pressure on the pure solvent to attain equilibrium has an advantage, namely that the second integral disappears so that only V° and not $\bar{V}$ need be known.

13-87 is an exact relation, but the data required to use it rigorously, especially values for $\bar{V}_A$ as a function of the pressure and of the concentration, are not usually available. In many cases this is no serious drawback because the concentrations and the osmotic pressures will be small, and good limiting-law approximations will serve. There is the drawback, however, that we may run into negative pressures, and there is a rather low limit to the tension that the cohesive forces in a column of liquid will support. Nevertheless the method developed by Frazer and Patrick[20] and refined by Williamson[21] has actually been used to measure osmotic pressures in dilute solutions by experiments in which there is a negative pressure on the solvent column (varying up to the order of 150 torr).

[20] I. Frazer and W. Patrick, *Z. physik. Chem.* **130,** 691 (1927).
[21] A. T. Williamson, *Proc. Roy. Soc.* **195,** 97 (1948).

PROBLEM 13-36 Show that if there are solutions on both sides of the osmotic membrane the osmotic equilibrium condition would be

$$RT \ln a_A'' - RT \ln a_A' = \int_{P^*}^{P'} \bar{V}_A' \, dP - \int_{P^*}^{P''} \bar{V}_A'' \, dP \qquad (13\text{-}88)$$

where a_A' and a_A'' are the activities of the solvent in the two solutions at P^*, P', and P'' are the equilibrium pressures (thus the osmotic pressure would be $P'' - P'$) and $\bar{V}_A'$ and $\bar{V}_A''$, which will be functions of P, are the solvent partial molal volumes in the two solutions.

::

PROBLEM 13-37 Show that if the pressures are reasonably small so that the compressibility of the liquids can be neglected, 13-87 becomes

$$RT \ln a_A^* = V_A^\circ(P^\circ - P^*) - \bar{V}_A(P - P^*) \qquad (13\text{-}89)$$

Show also that by the additional approximation that the solution is dilute enough so that $V_A^\circ = \bar{V}_A$ we get the simple result

$$RT \ln a_A = -\Pi V_A^\circ \qquad (13\text{-}90)$$

::

PROBLEM 13-38 We tentatively defined the ideal dilute solution as one in which the solvent obeyed Raoult's law and the solute obeyed Henry's law. Subsequently we showed that if Raoult's law (in terms of activities) was obeyed by the solvent, Henry's law must be obeyed by the solute. We can then redefine the ideal dilute solution as one for which $a_A = X_A$, and we would expect actual solutions to approach this behavior as $X_A \to 1$. Thus, for most cases if 13-90 is a good approximation then it is also a good approximation to set $a_A = X_A$. Using this (good) approximation calculate for 25 °C the osmotic pressure against pure water of a 0.001m sucrose solution. Give the answer in (a) atmospheres, (b) torrs, (c) centimeters of water.

Answer (c): 25.2 cm

::

PROBLEM 13-39 To evaluate $\ln a$ in Problem 13-38 you probably wrote $\ln (1 + x) = x - (x^2/2) + \cdots$ and discarded all but the first term of the series. Certainly this is a justifiable procedure for the example given. Show that for very dilute solutions a good approximation for the osmotic pressure is given by

$$\Pi = \frac{RT}{V_A^\circ} X_B \qquad (13\text{-}91)$$

and by

$$\Pi = \frac{n_B RT}{V} \qquad (13\text{-}92)$$

where B refers to solute and n_B is the number of moles of solute in a volume V of solution.

::

It is instructive to consider another way of deriving the osmotic equilibrium relationships. Figure 13-7 is our apparatus. It is more like usual osmometers than is the device shown in Figure 13-6. In the device shown in Figure 13-7 we have two barriers assumed to be selectively permeable to the solvent, namely, the membrane M and the vapor path above the liquids (we must here, then, require that the solute be nonvolatile). We assume that the vapor space contains only solvent vapor.

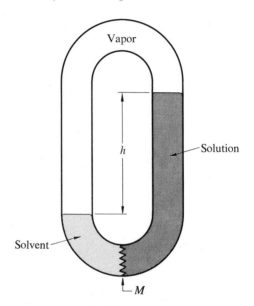

FIGURE 13-7

If we call P_g the vapor pressure of solvent A over the solution and P_g° the vapor pressure of A over the pure solvent, then for equilibrium $P_g^\circ - P_g$ must equal the pressure represented by a vapor column of height h, that is

$$\rho_g g h = P_g^\circ - P_g \tag{13-93}$$

where ρ_g is the average density of the vapor column. But we can write another expression for ρ_g, namely

$$\rho_g = \frac{n_A M_A}{V_g} \tag{13-94}$$

where n_A is the number of moles of A in a vapor volume V_g and M_A is the molecular weight of the vapor. Using the ideal gas approximation and assuming that $P_g^\circ - P_g \ll P_g^\circ$ we get

$$\rho_g = \frac{M_A P_g^\circ}{RT} \tag{13-95}$$

If we assume that the height of the surface of the pure solvent above the osmotic barrier is negligibly small we do not have to consider densities of both solvent and solution, and we can write for the osmotic pressure

$$\Pi = \rho_\ell g h \tag{13-96}$$

where ρ_ℓ is the density of the solution.

Purely formally we can write

$$\rho_\ell = \frac{M_A}{V_\ell} \tag{13-97}$$

where V_ℓ is the volume of solution which has the weight of one mole of solvent. Thus, V_ℓ will approach V_A as the solution becomes more and more dilute.

PROBLEM 13-40 By solving for h from 13-93 and 13-95 and combining the result with 13-96 and 13-97 show that the two relations shown in 13-98 are good approximations for dilute solutions and notice that the second of these is equivalent to 13-91

$$\frac{RT(P_g^\circ - P_g)}{V_\ell P_g^\circ} \approx \Pi \approx \frac{RT}{V_\ell} X_B \tag{13-98}$$

::

It should be clear that the equilibrium situation in the apparatus of Figure 13-7 would not be changed if M were made a completely impenetrable barrier. Therefore, such a device could be used to measure osmotic pressures without using a membrane. One might inquire, however, as to whether or not any difficulties would arise because of variations in density, and indeed of composition,[22] corresponding to the variation of the gravitational potential from place to place in the arms of the apparatus. The answer is that no important difficulties would arise. To arrive at this answer we have only to compare Figure 13-6 with Figure 13-7 and note that it is of no consequence to the portions of solution and solvent close to the dividing wall whether the pressures upon them come from columns of liquid or pistons; nor if they come from columns of liquid, what the compositions of these columns may be. Thus equations like 13-87 would apply rigorously to the properties of the solvent and solution at the bottom of the columns.

[22] The variation in composition might be appreciable if one were dealing with very large molecules.

PROBLEM 13-41   What could be said about the relative chemical potentials of the pure solvent and of the solvent in the solution at a level in the gravitational field other than at the bottom of the apparatus? Consider two vertical tubes, one containing pure solvent and the other a solution, joined throughout their entire length by a selective barrier permeable to solvent only.

▪▪

The answer to Problem 13-38 suggests that the osmotic pressure method should be an excellent one for getting activities and osmotic coefficients of solvents in very dilute solutions and then, through the Gibbs–Duhem relation, evaluating the activities or activity coefficients of the solutes. From the definition of the practical osmotic coefficient it follows that the limiting law corresponding to 13-90 is

$$\phi = \frac{1000 \Pi V_A^\circ}{M_A RT \nu m} \tag{13-99}$$

Experimental difficulties are sizeable, however. One of these is that of the suitably selective barrier. This problem can be satisfactorily solved in most cases by the use of the porous-disc, vapor barrier system used by Frazer and Patrick, Williamson and others. Another important problem is that of the extremely exact uniformity of temperature required. We can estimate how exacting this requirement is by further reference to Figure 13-7.

Suppose that we have in the apparatus pure water and an aqueous solution corresponding to an osmotic pressure of 10^{-3} atm (about 1.0 cm of water) at 25 °C. We can rearrange 13-98 to give the difference between the vapor pressure over the solution and that over the solvent as

$$P_g^\circ - P_g = \frac{V_\ell P_g^\circ \Pi}{RT} \tag{13-100}$$

Putting in the numbers this gives us for $P_g^\circ - P_g$ a value of about 2.2×10^{-8} atm.

Now suppose that we change the temperature at the surface of the solution so that the vapor pressure there is equal to P°. By the arguments used in getting 13-98 and making reasonable assumptions about the low thermal conductivity of the vapor we can conclude that the system will adjust itself to a steady state in which $h = 0$, i.e., in which the apparent (false) osmotic pressure is zero. How much change in temperature would be required to bring about this result? We can answer this question by the Clapeyron equation

$$dT = \frac{T \Delta V}{\Delta H} dP \tag{13-6}$$

Putting in the 2.2×10^{-8} atm which we have calculated for dP and the values for T, ΔV, and ΔH we get $dT \approx 1.2 \times 10^{-5}$ degrees. Thus, about twelve millionths of a degree lack of temperature uniformity could completely wipe out the osmotic equilibrium effect (or double it). It is probably because of this extreme temperature sensitivity that it has not been possible[23] to exploit as fully as could otherwise have been done the use of osmotic pressure measurements to study very dilute solutions.

The Donnan[24] equilibrium is a case of selective-barrier equilibrium closely related to osmotic pressure phenomena. The barrier in this case permits the passage of certain ions and prevents the passage of others, and it can be treated as either permeable (osmotic equilibrium) or impermeable (nonosmotic equilibrium) to the solvent. For dilute solutions the concentration distribution on the two sides of the membrane would be substantially the same for the osmotic and the nonosmotic cases.

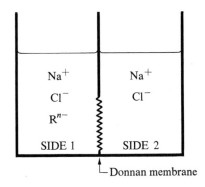

FIGURE 13-8

Figure 13-8 represents a nonosmotic Donnan equilibrium experiment. On SIDE 1 of the membrane is a solution containing dissolved NaCl and dissolved Na_nR. Na_nR is a strong electrolyte. The barrier is permeable to Na^+ and to Cl^- but not to R^{n-} (perhaps because of its size). The three main principles involved in deriving expressions for the equilibrium concentrations are simple. They are (1) constancy of the total amounts of the salts in the system, (2) equality of chemical potential (activity) of the substance which can get through the barrier (NaCl) on the two sides of the membrane, and (3) electrical neutrality. But the problem of activity coefficients can become very troublesome. As an instructive exercise we treat the system shown in Figure 13-8 on the basis that activity coefficients can be taken as 1, but we later see evidence that this may not be a very good approximation even in very dilute solutions.

[23] See Stuart A. Rice and Mitsuru Nagasawa, *Polyelectrolyte Solutions*, p. 393, Academic Press, 1961.

[24] F. G. Donnan, *Zeit. fur Electrochem.*, **17**, 572 (1911).

For a specified equilibrium concentration on SIDE 1 the equilibrium concentration on SIDE 2 will not depend on the volumes of solution on the two sides. We simplify our calculations, then, by assuming equal volumes. We use symbols such as (Na^+) to represent either activity or concentration, using the approximation mentioned in the preceding paragraph. Using C_0 to represent the average concentration of Na^+ we can write the following relations (the subscripts 1 and 2 refer to the respective sides of the membrane).

$$(Na^+)_1(Cl^-)_1 = (Na^+)_2(Cl^-)_2 \quad \begin{bmatrix} \text{Equality} \\ \text{of activity} \\ \text{of NaCl} \end{bmatrix} \qquad (13\text{-}101)$$

$$n(R^{n-}) + (Cl^-)_1 = (Na^+)_1 \quad \begin{bmatrix} \text{Electrical neutrality} \\ \text{on SIDE 1} \end{bmatrix} \qquad (13\text{-}102)$$

$$(Na^+)_2 = (Cl^-)_2 \quad \begin{bmatrix} \text{Electrical neutrality} \\ \text{on SIDE 2} \end{bmatrix} \qquad (13\text{-}103)$$

$$(Na^+)_1 + (Na^+)_2 = 2C_0 \qquad (13\text{-}104)$$

$$(Cl^-)_1 + (Cl^-)_2 = 2C_0 - n(R^{n-}) \qquad (13\text{-}105)$$

PROBLEM 13-42 Solve the preceding set of simultaneous equations to get all of the concentrations in terms of C_0 and (R^{n-}). Show that at the Donnan equilibrium, for example,

$$(Na^+)_1 = \frac{4C_0^2}{4C_0 - n(R^{n-})} \qquad (13\text{-}106)$$

and

$$\Delta C = \frac{(R^{n-})[4C_0 + n(n-1)(R^{n-})]}{4C_0 - n(R^{n-})} \qquad (13\text{-}107)$$

in which ΔC is the difference between the total ionic concentration on the two sides of the membrane.

∷

The expression for ΔC shown in 13-107 has some practical significance. Osmotic pressure experiments would measure ΔC. By considering the limiting cases $C_0 \gg n^2(R^{n-})$ and $C_0 \ll n(R^{n-})$ it can be seen that only in the first limiting case can osmotic pressure experiments alone be interpreted in terms of the molecular weight of a polyion, because only in this case is ΔC a function of (R^{n-}) not involving the (at the outset usually unknown) charge number n.

Nagasawa, Izumi, and Kagawa[25] studied a pair of double cells which we represent as equivalent to the double cell

Hg, Hg_2Cl_2, $NcCl_{(m_1)}$, $Na_nR_{(m_2)}$, Na-amalgam |

| Na-amalgam, $NaCl_{(m_3)}$, Hg_2Cl_2, Hg   (13-108)

in which Na_nR is the sodium salt of polyvinyl alcohol sulfuric acid.

[25] M. Nagasawa, M. Izumi, and I. Kagawa, *J. Polymer Sci.*, **37**, 375 (1959).

PROBLEM 13-43   Taking the far left electrode as anode consider the passage of 1 faraday through this double cell. Show that the net change in state is equivalent to the transfer of 1 mole of NaCl from the left-hand cell to the right-hand cell. Show that the ratio of the activities of NaCl in the two cells in terms of ε, the cell potential, is

$$\varepsilon = \frac{RT}{\mathscr{F}} \ln \frac{a_{(l)}}{a_{(r)}} \qquad (13\text{-}109)$$

Show that in terms of $\gamma_\pm$, the mean ion activity coefficient of NaCl in the solution containing the polyion, the relation is

$$\varepsilon = \frac{RT}{\mathscr{F}} \ln \frac{(m_1 + nm_2)m_1}{a_{(r)}} + \frac{2RT}{\mathscr{F}} \ln \gamma_\pm \qquad (13\text{-}110)$$

::

The value of $a_{(r)}$, the activity of NaCl in the right-hand cell, was known and kept constant, and various values were given to m_1 and to the equivalent concentration of polyion, nm_2. Hence it was possible through the emf measurements to determine $\gamma_\pm$ as a function of m_1 and nm_2. Some of the results of Nagasawa, Izumi, and Kagawa are reproduced in Figure 13-9. A striking feature of these graphs is that the presence of a given concentration of polyion depresses the activity coefficient of NaCl increasingly as the concentration of NaCl is decreased. It is evident that the approximations which lead to simple Debye–Hückel expressions for dilute solutions of small ions are inadequate for dealing with solutions containing polymeric ions which are large and have large charge numbers.

Selective barriers have proven useful in electrochemical cells. Delimarskii[26] and coworkers have studied fused-salt cells of the kind

$$\text{Na} \overset{B}{\,|\,} \text{NaCl, CuCl(melt)/Cu} \qquad (13\text{-}111)$$

in which B is a relatively thin glass barrier.

PROBLEM 13-44   If one considers the barrier B as a solid electrolyte permeable only to sodium ions show that per faraday the change in state for the cell is

$$\text{Na} + \text{CuCl} \rightarrow \text{NaCl} + \text{Cu} \qquad (13\text{-}112)$$

and that the standard Gibbs energy increment for reaction 13-112 would be related to the cell potential by

$$-\Delta G^\circ = \mathscr{F}\varepsilon + RT \ln \frac{a_{\text{NaCl}}}{a_{\text{CuCl}}} \qquad (13\text{-}113)$$

::

[26] I. U. K. Delimarskii and B. F. Markov, *Electrochemistry of Fused Salts*, (English translation), Andromeda Books, 2419 M Street, Washington, D.C., 1961.

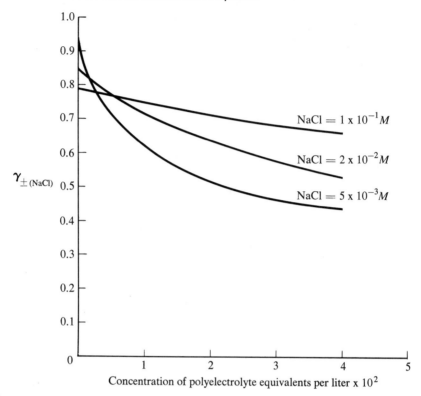

FIGURE 13-9

       The most familiar example of the use of a selective barrier in electrochemical cells is the glass electrode used in making pH measurements. It is difficult to give a generally-valid thermodynamic definition of pH, and the mechanism of operation of the glass membrane in aqueous (or nonaqueous) solutions is doubtless complex.[27] But the model of a selective solid-electrolyte barrier permeable only to hydrogen ions (in nonthermodynamic terms, a barrier in which the transference number of H^+ is one and the transference number of other ions is zero) predicts results approximating experimental facts found with certain glasses used in certain kinds of solutions.

       PROBLEM 13-45   The H^+-selective barrier model is represented below

$$Hg, \ Hg_2Cl_2(c), \ HCl_{m_1} \ \blacksquare \ HCl_{m_2}, \ Hg_2Cl_2, \ Hg \qquad (13\text{-}114)$$
$$\text{(glass)}$$

[27] See Roger G. Bates, *Determination of pH, Theory and Practice*, John Wiley and Sons, Inc., 1964.

Consider the fact that because only hydrogen ions can get through the glass all the current must be carried by them. Thus, for the passage of 1 faraday through the cell, 1 mole of H^+ must be transferred from one side of the glass to the other. Describe the net change in state corresponding to the passage of 1 faraday, taking the left electrode as anode. Show that the potential of the cell is

$$\varepsilon = \frac{RT}{\mathscr{F}} \ln \frac{a_1}{a_2} \tag{13-115}$$

in which a_1 is the activity of HCl at a concentration m_1 and a_2 at m_2.

::

EXCESS PROPERTIES

Partial molal properties of components of solutions are sometimes reported in terms of excess properties. For example, the excess chemical potential μ^E would be defined by

$$\mu^E = \mu - \mu^i \tag{13-116}$$

in which μ^i is the value μ would have if the solution were ideal. Let us first get an expression for μ^i. We can write

$$\mu_A = \left(\frac{\partial H}{\partial n_A}\right)_{T,P,n_B\cdots} - T\left(\frac{\partial S}{\partial n_A}\right)_{T,P,n_B\cdots} \tag{13-117}$$

PROBLEM 13-46 We derived the expression for the entropy of mixing in ideal solutions:

$$\Delta S_{\mathrm{mix}} = -R(n_A \ln X_A + n_B \ln X_B + \cdots) \tag{8-80}$$

We can write also

$$S = n_A S_A^\circ + n_B S_B^\circ + \cdots + \Delta S_{\mathrm{mix}} \tag{13-118}$$

By differentiating 13-118 with respect to n_A show that the partial molal entropy of A in an ideal solution is

$$\bar{S}_A = S_A^\circ - R \ln X_A \tag{13-119}$$

(S_A° and S_B° are the molal values for the pure components.)

::

PROBLEM 13-47 We can write an expression analogous to 13-118 for H and conclude that

$$\bar{H}_A = H_A^\circ + \left(\frac{\partial \Delta H_{\mathrm{mix}}}{\partial n_A}\right)_{T,P,n_B\cdots} \tag{13-120}$$

By combining 13-119 and 13-120 show that

$$\bar{G}_A - G_A^\circ = \left(\frac{\partial \Delta H_{mix}}{\partial n_A}\right)_{T,P,n_B\cdots} + RT \ln X_A \qquad (13\text{-}121)$$

We know that for an ideal solution X_A can always be set equal to the activity of A. We know further that by definition

$$\bar{G}_A - G_A^\circ = RT \ln a_A \qquad (13\text{-}122)$$

In view of the fact that ΔH_{mix} will be zero in any kind of solution under certain circumstances, that is, when we add A to a solution in which $X_A \to 1$ or B to one in which $X_B \to 1$, we can conclude by a comparison of 13-121 and 13-122 that for an ideal solution

$$\Delta H_{mix} = 0 \quad \begin{bmatrix} \text{Ideal} \\ \text{solution} \end{bmatrix} \qquad (13\text{-}123)$$

::

From 13-121 and the conclusion in the preceding problem we can write for $\mu_A{}^i$

$$\mu_A{}^i = \mu_A^\circ + RT \ln X_A \qquad (13\text{-}124)$$

or

$$\mu_A{}^i - \mu_A^\circ = RT \ln X_A \qquad (13\text{-}125)$$

We can also conclude that because

$$\mu_A - \mu_A^\circ = \left(\frac{\partial \Delta H_{mix}}{\partial n_A}\right)_{T,P,n_B\cdots} - T\left(\frac{\partial \Delta S_{mix}}{\partial n_A}\right)_{T,P,n_B\cdots} \qquad (13\text{-}126)$$

then

$$\mu_A - \mu_A{}^i \equiv \mu^E = \left(\frac{\partial \Delta H_{mix}}{\partial n_A}\right)_{T,P,n_B\cdots} - T\left(\frac{\partial \Delta S_{mix}}{\partial n_A}\right)_{T,P,n_B\cdots} - RT \ln X_A \qquad (13\text{-}127)$$

It is evident that if 8-80 is taken as the definition of ideal entropy of mixing, then if the entropy of mixing is ideal in a particular case, the last two terms in 13-127 cancel each other and for that case

$$\mu^E = \left(\frac{\partial \Delta H_{mix}}{\partial n_A}\right)_{T,P,n_B\cdots} \qquad (13\text{-}128)$$

Hildebrand[28] proposed the name regular solutions for solutions in which the entropy of mixing is ideal but the heat of mixing is not zero. One would expect to find this behavior in solutions in which the intermolecular forces between the different kind of molecules were different but not sufficiently different to prevent the thermal translational energies of the molecules from bringing about a nearly completely random distribution. The expression for

[28] J. H. Hildebrand, *J. Am. Chem. Soc.*, **51**, 66 (1929).

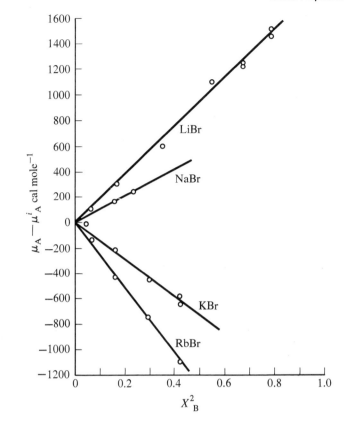

FIGURE 13-10 Excess free energy of AgBr in fused alkali bromide solutions. A refers to AgBr, B to the indicated alkali bromide.

ΔH_{mix} obtained through statistical considerations indicate that for certain types of regular solutions (symmetric) the expressions for μ^E in a binary solution of A and B are

$$\mu_A^{E} = kX_B^{2} \qquad (13\text{-}129)$$

$$\mu_B^{E} = kX_A^{2} \qquad (13\text{-}130)$$

Figure 13-10 shows results of Hildebrand and Salstrom[29] on solutions of silver bromide in various alkali bromides at 550° and 600 °C. The curves (lines) for these two temperatures are indistinguishable and indicate that Equation 13-129 is closely followed by AgCl (taken as component A) in these solutions.

[29] Joel H. Hildebrand and Edward J. Salstrom, *J. Am. Chem. Soc.*, **54**, 4257 (1932).

PROBLEM 13-48 From the data in Figure 13-10 calculate the activity coefficient of AgBr in a 0.5 mole fraction solution in LiBr at 500 °C. Calculate also $\bar{L}_{AgBr}$, the relative partial molal enthalpy of AgBr in this solution.

::

We began this book with a quotation from James. We terminate it rather arbitrarily at about this point with John's[30] hyperbole

And there are also many other things . . . . which if they should be written every one, I suppose the world itself would not contain the books that should be written.

ADDITIONAL PROBLEMS

PROBLEM 13-49 An *extensive* property can be defined as one whose value is n times as great for a collection of n identical systems as it is for any one of those systems. An *intensive* property is one which has the same value for the collection of systems as for the one system. Classify the following as extensive or intensive properties: P, V, T, E, S, A, G, viscosity, surface tension, surface energy, mole fraction.

::

PROBLEM 13-50 Mathematically a homogeneous function, f, of the kth degree is one which satisfies the equation:

$$f(nx, ny, \ldots) = n^k f(x, y, \ldots) \tag{13-131}$$

Show that the extensive properties of a solution are homogeneous functions of the first degree in the number of moles of the components of the solution.

::

PROBLEM 13-51 By differentiating 13-131 partially with respect to n show that:

$$x \frac{\partial f}{\partial(nx)} + y \frac{\partial f}{\partial(ny)} + \cdots = kf \tag{13-132}$$

This must hold for any value of n. Hence show that

$$x \frac{\partial f}{\partial x} + y \frac{\partial f}{\partial y} + \cdots = kf \tag{13-133}$$

This is Euler's theorem.

::

PROBLEM 13-52 Show, in the light of Problems 13-49, 13-50, and 13-51 that 13-25

$$V = n_1 \bar{V}_1 + n_2 \bar{V}_2 \tag{13-25}$$

follows directly from Euler's theorem.

::

[30] The Gospel according to John 21:25.

PROBLEM 13-53   By noting that at constant T and P

$$dV = \bar{V}_1 \, dn_1 + \bar{V}_2 \, dn_2 + \cdots \qquad (13\text{-}23)$$

and by combining 13-23 with the expression for dV obtained by differenting 13-25, derive the Gibbs–Duhem relation

$$n_1 \, d\bar{V}_1 + n_2 \, d\bar{V}_2 + \cdots = 0 \qquad (13\text{-}134)$$

Other Gibbs–Duhem equations such as

$$X_A \, d\mu_A + X_B \, d\mu_B = 0 \qquad (12\text{-}9)$$

can be derived likewise.

Appendix A

Table 10-2

Formula	ΔH_0°	ΔH_{298}°	ΔG_{298}°	$H_{298}^\circ - H_0^\circ$	S°	C_P°
O_2(g)	0	0	0	2.0746	48.996	7.016
H_2(g)	0	0	0	2.0238	31.208	6.889
H_2O(l)		−68.315	−56.688		16.71	17.995
H_2O(g)	−57.102	−57.796	−54.635	2.3667	45.104	8.025
H_2O_2(l)		−44.88	−28.78		26.2	21.3
H_2O_2(aq)		−45.69	−32.05		34.4	
He(g)	0	0	0	1.481	30.1244	4.9679
Cl_2(g)	0	0	0	2.193	53.288	8.104
Cl_2(aq)		−5.6	1.65		29.	
HCl(g)	−22.020	−22.062	−22.774	2.066	44.646	6.96
HCl(aq)		−39.952	−31.372		13.5	−32.6
Br_2(l)		0	0	5.859	36.384	18.090
Br_2(g)	10.921	7.387	0.749	2.324	58.647	8.62
Br_2(aq)		−0.62	0.94		31.2	
S (rhombic)	0	0	0	1.054	7.60	5.41
S (monoclinic)		0.08				
SO_2(g)	−70.336	−70.944	−71.749	2.521	59.30	9.53
H_2SO_4(l)		−194.548	−164.942	6.748	37.501	33.20
H_2SO_4(aq)		−217.32	−177.97		4.8	−70.
NO_2(g)	276.6	277.4				
N_2O_4(g)	4.49	2.19	23.38	3.918	72.70	18.47
NH_3(g)	−9.34	−11.02	−3.94	2.388	45.97	8.38
NH_3(aq)		−19.19	−6.35		26.6	
HN_3(g)	71.82	70.3	78.4	2.599	57.09	10.44
C (graphite)	0	0	0	0.251	1.372	2.038
C (diamond)	0.5797	0.4533	0.6930	0.125	0.568	1.4615
CO_2(g)	−93.964	−94.051	−94.258	2.2378	51.06	8.87
CO_2(aq)		−98.90	−92.26		28.1	
CH_4(g)	−15.970	−17.88	−12.13	2.388	44.492	8.439
CH_3OH(l)		−57.04	−39.76		30.3	19.5
CH_3OH(g)	−45.355	−47.96	−38.72	2.731	57.29	10.49
Ag(c)	0	0	0		10.20	6.10
Ag_2O(c)		−7.42	−2.68		29.0	
AgCl(c)		−30.370	−26.244		23.0	
Na(c)	0	0	0		12.24	6.75
NaOH(aq)		−112.236	−100.184		11.9	
NaCl(c)	−97.755	−98.232	−91.785	67.2771	17.30	11.88
NaCl(aq)		−97.302	−93.939		27.6	
Hg_2Cl_2(c)		−63.32	−50.350		46.8	24.3
CuCl(c)		−32.2	−28.4		21.9	

Table 10-2 lists the standard Gibbs energy of formation at 25 °C and other thermodynamic properties of some elements and compounds. These values are taken from National Bureau of Standards Circular 500. "Selected Values of Chemical Thermodynamic Properties" by Frederick D. Rossini, Donald Wagman, William H. Evans, Samuel Levine, and Irving Jaffe (1952) and from the first two parts of the revision of this work as NBS Technical Notes 270-1 and 270-2 by D. D. Wagman, W. H. Evans, I. Halow, V. B. Parker, S. M. Bailey, and R. H. Schumm (1966). ΔH_0°, ΔH_{298}°, ΔG_{298}°, $H_{298}^\circ - H_0^\circ$ are in kcal mole^{-1}; S° and C_P° are in cal deg^{-1} mole^{-1}. The subscript (aq) refers to an aqueous solution, the standard state in these cases being defined on the basis of Equation 10-21. The subscripts (g), (l), and (c) mean gas, liquid, and crystal.

Appendix B

The data in Tables 10-3A, 10-3B, 10-3C are taken from tables in *Thermodynamics* by Lewis and Randall, revised by Pitzer and Brewer, McGraw-Hill Book Company, 1961. They are used by permission of the McGraw-Hill Book Company, Inc.

A very large compilation of data of the kind contained in Appendix B is the *JANAF Thermochemical Tables* (1965) and the Addendum (1966) and Second Addendum (1967) and Third Addendum (1968). These tables were prepared under contract by the staff of the Thermal Research Laboratory of the Dow Chemical Company, D. R. Stull, Project Director, technically

Table 10-3A Free Energies Based on H°_{298}

(Solid and liquid substances)

	$-(G^{\circ} - H^{\circ}_{298})/T$, cal deg^{-1}					ΔH°_{298} kcal	$H^{\circ}_{298} - H^{\circ}_0$ kcal
	298.15 °K	500 °K	1000 °K	1500 °K	2000 °K		
Ag	10.20	10.90	13.40	15.78	(17.99)		1.373
Br$_2$	36.4						3.24
C	1.37	1.67	3.04	4.37	5.51		0.251
Ca	9.95	10.69	13.59	(16.53)			1.38
Cu	7.97	8.64	11.09	13.33	(15.47)		1.201
Hg	18.17	18.92					2.23
Na	12.24	13.49	16.96				1.531
S	7.62	19.6					1.79
AgCl	23.00	24.49	31.25	(37)		−30.36	
CaCl$_2$	27.2	29.3	36.6	44.6		−190.0	
CuCl	20.8	22.2	29.3			−32.6	
Hg$_2$Cl$_2$	46.2	(48)				−63.3	
HgCl$_2$	34.5	36.5				−53.4 ± 1	
CuBr	23.0	(24)	(31)			−25.3	
NaCl	17.33	18.75	23.85			−98.2	
Ag$_2$O	29.1	30.99				−7.2	
Ag$_2$S	33.5	35.93	45.24			−7.6	
Cu$_2$O	22.4	24.33	31.48	(37.2)		−40.4	
CuO	10.19	11.43	16.09	(20.07)		−37.6	
CaO	9.5	10.73	15.26	18.94	21.9	−151.79	

assisted by the Joint Army–Navy–Air Force–ARPA–NASA Thermonuclear Working Group. They are U.S. Government publications and are distributed by *Clearing House* for Federal Scientific and Technical Information National Bureau of Standards, Washington, D.C.

Table 10-3B   Free Energies Based on H_{298}°

(Gaseous substances)

	\multicolumn{5}{c}{$-(G^\circ - H_{298}^\circ)/,T$ cal deg^{-1}}	$H_{298}^\circ - H_0^\circ$ kcal	ΔH_{298}° kcal
	298.15 °K	500 °K	1000 °K	1500 °K	2000 °K		
Ag	44.32	41.89	43.85	45.37	46.56	1.481	68.4
Br	41.81	42.37	44.34	45.88	47.11	1.481	26.76
Br$_2$	65.40	66.40	69.92	72.65	74.77	2.453	52.5 ± 1
C	37.76	38.33	40.30	41.81	43.00	1.56	170.9
Cl	39.46	40.06	42.19	43.83	45.09	1.499	28.94
Cl$_2$	53.29	54.24	57.63	60.32	62.43	2.194	0
H$_2$	31.21	32.00	34.76	36.95	38.69	2.024	0
Na	36.71	37.28	39.25	40.76	41.95	1.481	25.9
O	38.47	39.06	41.07	42.61	43.81	1.607	59.55
O$_2$	49.01	49.83	52.78	55.19	57.15	2.075	0
S$_2$	54.51	55.42	58.72	61.35	63.43	2.141	30.84
CuCl	56.5	57.46	60.9	...	65.7		32
CuBr	59.2	60.18	63.66	...	68.5		38
HgCl	62.1	63.09	66.57	...	71.4		19
HgCl$_2$	70.3	71.9	77.6	...	...		−33.5 ± 1

Table 10-3C   Free Energies Based on H_0°

(Gaseous substances)

	\multicolumn{5}{c}{$-(G^\circ - H_0^\circ)/T$, cal deg^{-1}}	ΔH_{298}° kcal	$H_{298}^\circ - H_0^\circ$ kcal	ΔH_0° kcal
	298 °K	500 °K	1000 °K	1500 °K	2000 °K			
Br$_2$	50.85	54.99	60.80	64.31	66.83		2.325	8.37
Cl$_2$	45.93	49.85	55.43	58.85	61.34		2.194	0
H$_2$	24.42	27.95	32.74	35.59	37.67		2.024	0
O$_2$	42.06	45.68	50.70	53.81	56.10		2.07	0
HCl	37.72	41.31	46.16	49.08	51.23	−22.063	2.065	−22.019
HBr	40.53	44.12	48.99	51.95	54.13	−8.66	2.067	−8.1
H$_2$O	37.17	41.29	47.01	50.60	53.32	−57.798	2.368	−57.107
H$_2$O$_2$	46.95	51.72	59.15	64.28		−32.53	2.59	−31.04
NaCl	47.18	51.25	57.05	60.56	63.09	−42.7	2.295	−42.4

Name Index

Allmand, A. J., 159
Anderson, P. D., 96
Andrews, D. H., 200
Andrews, F. C., 104
Arrhenius, S. A., 161
Avogadro, A., 11

Bailey, S. M., 125, 207
Bates, R. G., 228
Beckett, C. W., 152
Benedict, W. S., 152
Berthelot, D., 98
Biron, E., 211
Bjerrum, N., 168, 184
Blake, F. C., 216
Boanerges, J., 232
Boltzmann, L., 101, 108, 109
Boyle, R., 9
Bridgman, P. W., 77
Brewer, L., 77, 88, 99, 125, 131, 156, 205
Brønsted, J. N., 168, 170
Brown, P. M. G., 218

Carnot, S., 25, 26, 27, 33, 34, 35, 40
Carroll, B., 77
Charles, J., 9
Clapeyron, E., 196, 197
Clausius, R. J. E., 51, 53, 161, 197
Clusius, K., 95
Cowan, C. L., 7
Craig, D. N., 208, 212
Curie, P., 89

Davies, C. W., 169
Debye, P., 8, 95, 168, 187
De Donder, T., 61, 134
Defay, R., 61

Delimarskii, Iu. K., 227
Dole, M., 131
Donnan, F. G., 225
Dorsey, N. E., 198
Duhem, P., 179, 180, 186, 233

Eastman, E. D., 95, 127
Ehlers, R. W., 157, 171
Ellis, J. H., 157
Epstein, P. S., 88
Eucken, A., 98
Euler, L., 232
Evans, W. H., 23, 125, 207
Everett, D. H., 61

Fano, L., 152
Faraday, M., 42
Fermi, E., 7
Frank, H. S., 162
Frazer, I., 220, 224

Gay-Lussac, J. L., 9
Gendrano, M. C., 154
Gerke, R. H., 98
Giauque, W. F., 97, 117, 130
Gibbs, J. W., 43, 65, 66, 68, 94, 179, 180, 186, 194, 214, 233
Glassner, A., 24
Gosh, J. C., 168
Greenewalt, C. H., 216
Guggenheim, E. A., 61, 101, 169

Halow, I., 125, 207
Hamer, W. J., 43, 183, 188, 194
Harned, H. S., 157, 171, 175, 183
Harteck, P., 95

Subject Index